Intermediate 2
Physics

2001 Exam

2002 Exam

2003 Exam

2004 Exam

2005 Exam

Leckie×Leckie

© **Scottish Qualifications Authority**

All rights reserved. Copying prohibited. No part of this publication may be reproduced, stored in a retrieval system, or transmitted in any form or by any means, electronic, mechanical, photocopying, recording or otherwise.

First exam published in 2001.
Published by Leckie & Leckie, 8 Whitehill Terrace, St. Andrews, Scotland KY16 8RN tel: 01334 475656 fax: 01334 477392
enquiries@leckieandleckie.co.uk www.leckieandleckie.co.uk

ISBN 1-84372-332-8

A CIP Catalogue record for this book is available from the British Library.

Printed in Scotland by Scotprint.

Leckie & Leckie is a division of Granada Learning Limited, part of ITV plc.

Acknowledgements

Leckie & Leckie is grateful to the copyright holders, as credited at the back of the book, for permission to use their material.
Every effort has been made to trace the copyright holders and to obtain their permission for the use of copyright material.
Leckie & Leckie will gladly receive information enabling them to rectify any error or omission in subsequent editions.

2001 | Intermediate 2

X069/201

NATIONAL
QUALIFICATIONS
2001

MONDAY, 4 JUNE
9.00 AM – 11.00 AM

PHYSICS
INTERMEDIATE 2

Read Carefully

1 All questions should be attempted.

Section A (questions 1 to 20)

2 Check that the answer sheet is for Physics Intermediate 2 (Section A).
3 Answer the questions numbered 1 to 20 on the answer sheet provided.
4 Fill in the details required on the answer sheet.
5 Rough working, if required, should be done only on this question paper, or on the first two pages of the answer book provided—**not** on the answer sheet.
6 For each of the questions 1 to 20 there is only **one** correct answer and each is worth 1 mark.
7 Instructions as to how to record your answers to questions 1–20 are given on page two.

Section B (questions 21 to 31)

8 Answer the questions numbered 21 to 31 in the answer book provided.
9 Fill in the details on the front of the answer book.
10 Enter the question number clearly in the margin of the answer book beside each of your answers to questions 21 to 31.
11 Care should be taken to give an appropriate number of significant figures in the final answers to calculations.

SECTION A

For questions 1 to 20 in this section of the paper, an answer is recorded on the answer sheet by indicating the choice A, B, C, D or E by a stroke made in ink in the appropriate box of the answer sheet—see the example below.

EXAMPLE

The energy unit measured by the electricity meter in your home is the

    A ampere

    B kilowatt-hour

    C watt

    D coulomb

    E volt.

The correct answer to the question is B—kilowatt-hour. Record your answer by drawing a heavy vertical line joining the two dots in the appropriate box on your answer sheet in the column of boxes headed B. The entry on your answer sheet would now look like this:

If after you have recorded your answer you decide that you have made an error and wish to make a change, you should cancel the original answer and put a vertical stroke in the box you now consider to be correct. Thus, if you want to change an answer D to an answer B, your answer sheet would look like this:

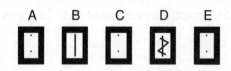

If you want to change back to an answer which has already been scored out, you should enter a tick (✓) to the RIGHT of the box of your choice, thus:

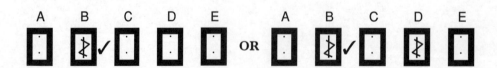

SECTION A

Answer questions 1–20 on the answer sheet.

1. Which of the following pairs contain two scalar quantities?

   A Force and mass
   B Weight and mass
   C Displacement and speed
   D Distance and speed
   E Displacement and velocity

2. Four tugs apply forces to an oil-rig as shown.

   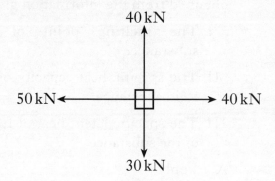

   Which of the following could represent the resultant force?

   A

   B

   C

   D

   E

3. Two identical balls X and Y are projected horizontally from the edge of a cliff. The paths they take are as shown below.

   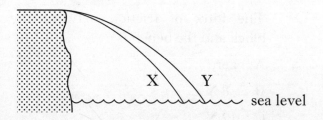

   A student made the following statements about the motion of the two balls.

   I   The balls take the same time to reach sea level.
   II The balls have the same vertical acceleration.
   III The balls have the same horizontal velocity.

   Which of these statements is/are correct?

   A I only
   B II only
   C I and II only
   D I and III only
   E II and III only

   [Turn over

4. A block of mass 3 kg is pulled across a horizontal bench by a force of 20 N as shown below.

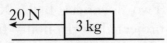

The block accelerates at 4 m/s².

The force of friction between the block and the bench is

A zero

B 8 N

C 12 N

D 20 N

E 32 N.

5. A ball of mass 2 kg rolls with a velocity of 4 m/s along a horizontal surface.

Which line of the table below correctly shows the momentum and kinetic energy of the ball?

	Momentum (kg m/s)	Kinetic energy (J)
A	2	4
B	4	8
C	4	16
D	8	8
E	8	16

6. A heater rated at 500 W is used to heat 1 kg of a substance. Initially the substance is in the solid state.

The following graph of temperature of substance against time is obtained.

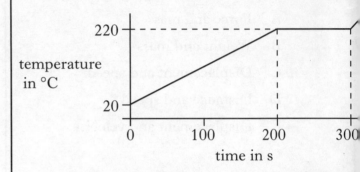

Which of the following values can be obtained from the information given?

I The melting point of the substance.

II The specific heat capacity of the solid substance.

III The specific latent heat of fusion of the substance.

A I only

B I and II only

C I and III only

D II and III only

E I, II and III

7. Which of the following is the unit of gravitational field strength?

A kg m/s

B N m

C N m/s

D kg m/s²

E N/kg

8. An electric motor with an input power of 1 kW is 80% efficient. The "wasted" energy is all transferred as heat energy. How much heat energy is produced in 1 s?

A 200 J
B 800 J
C 1000 J
D 2000 J
E 8000 J

9. A circuit is set up as shown below.

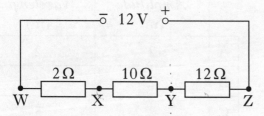

The potential difference between X and Y is

A 1·2 V
B 4·0 V
C 5·0 V
D 10·0 V
E 12·0 V.

10. A car headlamp is operating at its rated values of 12 V and 48 W.

Which of the following statements is/are correct?

 I The lamp uses energy at the rate of 48 joules per second.
 II The current through the lamp is 4 amperes.
 III 12 coulombs of charge flow through the lamp every second.

A I only
B II only
C I and II only
D II and III only
E I, II and III

11. The filament of a lamp has a resistance of 3 Ω and the current through the filament is 2 A.

The electrical power produced by the lamp is

A 1·5 W
B 6 W
C 12 W
D 18 W
E 36 W.

12. The input to an amplifier is 2 V a.c. at a frequency of 200 Hz. The amplifier has a gain of 8.

Which line in the table below correctly shows the output voltage and the output frequency?

	Output voltage (V)	Output frequency (Hz)
A	10	200
B	10	208
C	10	1600
D	16	200
E	16	1600

13. Which resistor in the diagram below has the smallest resistance?

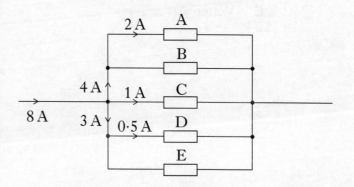

14. Which of the following devices converts heat energy into electrical energy?

    A Solar cell

    B Resistor

    C Thermocouple

    D Transformer

    E Transistor

15. The diagram shows part of the electromagnetic spectrum.

    | Radio waves | Micro-waves | Z | Visible light |

    The radiation in the region marked Z is called

    A ultraviolet

    B infrared

    C X-rays

    D sound

    E gamma rays.

16. Which of the following waves is a longitudinal wave?

    A Microwaves

    B Radio waves

    C Sound waves

    D Light waves

    E Water waves

17. The following diagram gives information about a wave.

    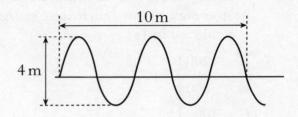

    Which line in the table below correctly shows the amplitude and wavelength of the wave?

	Amplitude (m)	Wavelength (m)
A	2	2
B	2	4
C	2	5
D	4	2
E	4	4

18. A ray of light passes from air into glass as shown.

    Which letter marks the angle of refraction?

    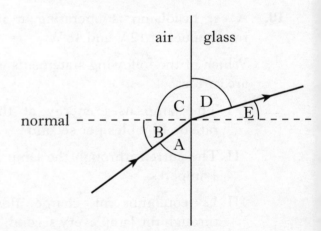

19. Measurements of the count rate from a radioactive source were taken using the apparatus shown below.

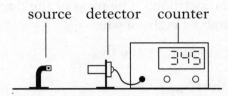

A sheet of paper, 2 mm of aluminium and 15 mm of lead were placed in turn between the radioactive source and the detector.

Information about the count rate obtained with and without the absorbers is given in the following table.

Absorber	Corrected count-rate (counts per second)
none	80
1 sheet of paper	65
2 mm of aluminium	35
15 mm of lead	5

The radiation emitted by the source is

A α only

B β only

C α and β only

D β and γ only

E α, β and γ.

20. Which row in the table below shows the correct units for activity and dose equivalent?

	Activity	Dose equivalent
A	becquerel	gray
B	becquerel	sievert
C	gray	sievert
D	gray	becquerel
E	sievert	gray

[Turn over

SECTION B

Write your answers to questions 21–31 in the answer book.

21. A flag is raised at the opening of an athletics competition. The mass of the flag is 0·5 kg and it is raised at constant speed through a height of 6 m.

    (a) Calculate the gravitational potential energy gained by the flag. — 2

    (b) A constant force of 7 N is applied to raise the flag.
    Calculate the work done raising the flag. — 2

    (c) Explain why there is a difference between the answers to parts (a) and (b). — 2

    (6)

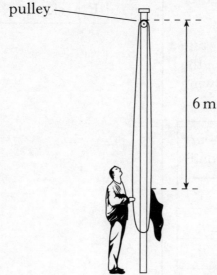

22. A rowing team is taking part in a race on calm water.

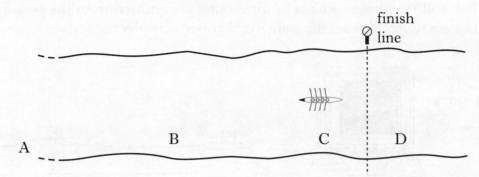

The following graph shows how it is predicted that the speed of the boat will vary with time during the stages A, B, C and D of the race.

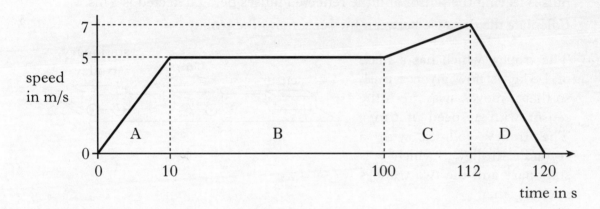

The prediction assumes that the frictional force on the team's boat remains constant at 800 N during the race.

(a) (i) State the size of the forward force applied by the oars during stage B. **1**

(ii) Calculate the acceleration of the boat during stage C. **2**

(iii) The total mass of the boat and its crew is 400 kg.
Calculate the size of the forward force applied by the oars during stage C. **3**

(iv) The boat crosses the finishing line after 112 seconds.
Calculate the distance the boat travels **from the instant it crosses the line** until it comes to rest. **2**

(b) The frictional force acting on the boat during stage D actually becomes smaller as the speed decreases.

(i) What will be the effect of this smaller frictional force on the time taken for the boat to come to rest? **1**

(ii) Sketch a graph of speed against time for stage D, assuming that the frictional force becomes smaller as the speed decreases. **1**

(10)

23. A sensor linked to a computer can be used to measure the distance between a trolley and the sensor. Pulses of ultrasound are emitted from the sensor. The pulses are reflected from the trolley and are detected by the sensor.

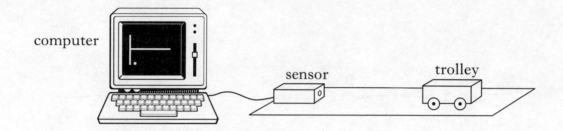

(a) Ultrasound travels at a speed of 340 m/s in air. The time between the pulses leaving the sensor and the reflected pulses being detected is 5 ms. Calculate the distance between the sensor and the trolley.

(b) The trolley, which has a mass of 1·5 kg, is now given a push so that it moves away from the sensor with a speed of 6 m/s. The trolley collides with a second trolley which is stationary and the two trolleys stick together.

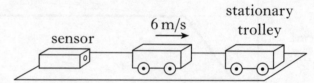

The computer produces the following speed-time graph of the motion before and after the collision.

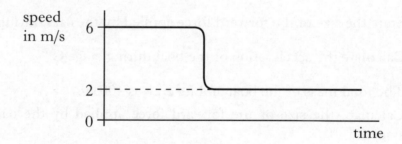

Calculate the mass of the second trolley.

24. Some bottles of water are placed in a compartment of a refrigerator.

The refrigerator reduces the temperature of the water from 22·0 °C to 10·0 °C.

The **total** mass of water in the bottles is 2·40 kg.

(a) The specific heat capacity of the water is 4200 J/kg °C.

Show that the heat energy lost by the water is 121 kJ, correct to 3 significant figures.

(b) The refrigeration system removes heat energy from the compartment at a rate of 100 J/s.

(i) Assuming that heat is removed **from the water** at this rate, how long will it take to lower the water temperature from 22·0 °C to 10·0 °C?

(ii) Explain why the actual time taken to lower the temperature of the water will be longer than the value you calculated in part (i).

25. A hotel owner decides to instal three lamps on the drive between the hotel and the street.

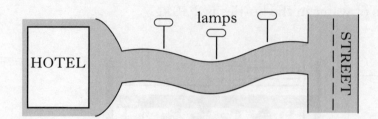

The circuit diagram below shows how the lamps are connected to the mains supply.

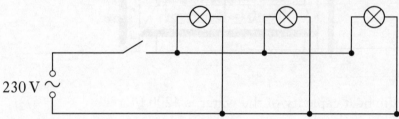

Each lamp has a rating of 230 V, 200 W.

(a) Explain why the lamps must be connected in parallel. **1**

(b) Calculate the resistance of each lamp. **2**

(c) Calculate the current drawn from the supply when all three lamps are operating. **3**

(d) The lamps are connected to the circuit shown below so that they come on automatically when it gets dark.

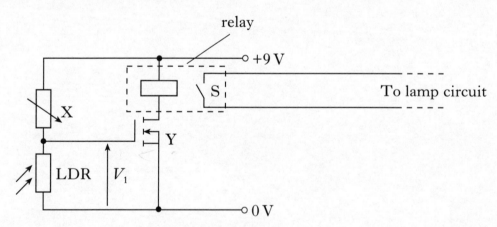

(i) Identify components labelled X and Y. **2**

(ii) Component Y switches on when the voltage V_1 reaches 2·4 V.
Switch S is closed when there is a current in the relay.
Explain how this circuit will switch the lamps on when it becomes dark. **3**

(11)

26. In a car battery charger, a transformer is used to step voltage down from 230 V to 12 V. The stepped down voltage is converted to d.c. using a converter. The circuit is shown below.

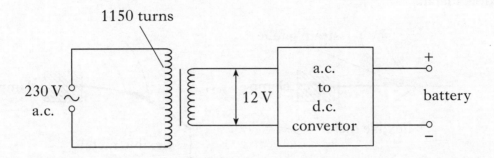

(a) There are 1150 turns on the primary coil.

Calculate the number of turns on the secondary coil. **2**

(b) (i) What do the initials d.c. stand for? **1**

(ii) Explain what d.c. means in terms of electron flow in a circuit. **1**

(c) The charger delivers a current of 400 mA to the battery for a period of 5 hours.

Calculate the charge delivered to the battery during this time. **2**

(6)

[Turn over

27. A strain gauge is an electrical device that is attached to an object.

    The strain gauge detects a change in the shape of the object.

    In the following diagrams, the strain gauge is shown attached to a piece of flexible metal.

    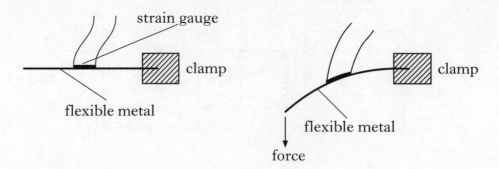

    When a force is applied to the end of the piece of metal, it bends.

    When the metal is bent, the strain gauge also bends and its resistance changes.

    The strain gauge is connected in series with a resistor, R, and a 9 V supply as shown in the circuit diagram below.

    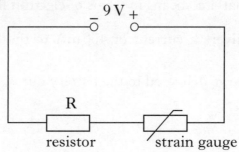

    (a) A student is asked to find the resistance of the strain gauge using a voltmeter and an ammeter.

    Redraw the diagram to show how the student should connect the meters to measure the resistance of the strain gauge. **2**

    (b) The student obtains the following results.

	Voltmeter reading (V)	Ammeter reading (mA)
No force applied	7·20	60·0
Force applied	7·23	59·0

    Does the resistance of the strain gauge increase or decrease when the force is applied to the piece of metal? You must justify your answer. **3**

    (c) Calculate the resistance of the resistor R. **3**

    (8)

28. A telecommunications company uses microwaves to transmit information between two positions A and B separated by a range of hills. A relay station on top of the hills receives and transmits the signals using curved reflectors.

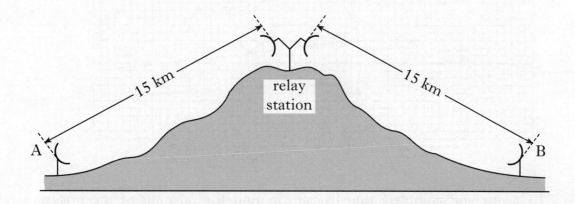

(a) Explain why a curved reflector is used to receive a signal. Your answer should include a diagram. **2**

(b) The microwaves have a wavelength of 15 mm and a speed of 3×10^8 m/s in air.
Calculate the frequency of the microwaves. **2**

(c) Calculate the minimum time taken by the microwaves to travel from A to B. **3**

(d) The relay station requires an energy source but is too remote to have a mains electricity supply. Suggest a possible alternative supply. **1**

(8)

[Turn over

29. (a) The diagram shows the path of one ray of light from the top of an object placed in front of a converging lens.

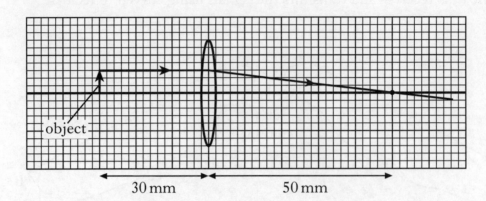

(i) Copy and complete the diagram to find the position of the image. Draw the image on your diagram.

You may use the graph paper provided. **2**

(ii) Using information from the diagram, calculate the power of the lens. **3**

(b) People with long sight need converging lenses to improve their vision. What is meant by long sight? **1**

(6)

30. The oil industry uses radioactive sources to monitor the flow of liquids in pipes. The complete detection system is attached to the outside of the pipe as shown.

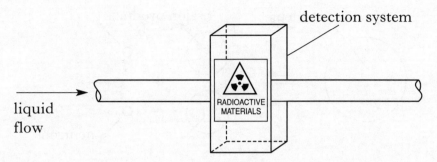

(a) The source used has an activity of 1·11 GBq.

Explain what is meant by this statement. 2

(b) A sample of tissue exposed to this radiation receives an absorbed dose of 0·13 mGy.

The quality factor of the radiation is 9. Calculate the dose equivalent for this sample. 2

(c) The system is surrounded by a large cage as shown in the diagram.

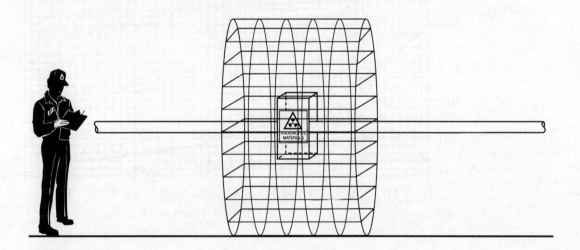

What is the purpose of this cage? 2

(6)

[Turn over

31. In a nuclear reactor, uranium nuclei are bombarded by neutrons. Fission products and additional neutrons are produced. Energy is also released.

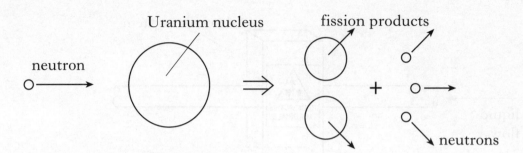

(a) In a reactor, what is the purpose of

    (i) the coolant? **1**

    (ii) the moderator? **1**

(b) Explain how the additional neutrons can cause a chain reaction. **2**

(c) A graph of activity against time for a sample of one of the fission products is shown below.

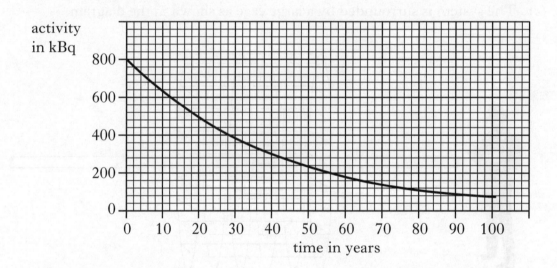

    (i) From the graph, determine the half life of the fission product. **2**

    (ii) A scientist states that the sample will be safe only when the activity falls to 120 kBq. How long will it take for the activity to fall to this level? **1**

    (iii) State a suitable method of storing the sample during the time it takes for the activity to fall to the safe level. **1**

(8)

[END OF QUESTION PAPER]

2002 | Intermediate 2

X069/201

NATIONAL
QUALIFICATIONS
2002

WEDNESDAY, 22 MAY
1.00 PM – 3.00 PM

PHYSICS
INTERMEDIATE 2

Read Carefully

1 All questions should be attempted.

Section A (questions 1 to 20)

2 Check that the answer sheet is for Physics Intermediate 2 (Section A).
3 Answer the questions numbered 1 to 20 on the answer sheet provided.
4 Fill in the details required on the answer sheet.
5 Rough working, if required, should be done only on this question paper, or on the first two pages of the answer book provided—**not** on the answer sheet.
6 For each of the questions 1 to 20 there is only **one** correct answer and each is worth 1 mark.
7 Instructions as to how to record your answers to questions 1–20 are given on page two.

Section B (questions 21 to 31)

8 Answer the questions numbered 21 to 31 in the answer book provided.
9 Fill in the details on the front of the answer book.
10 Enter the question number clearly in the margin of the answer book beside each of your answers to questions 21 to 31.
11 Care should be taken to give an appropriate number of significant figures in the final answers to calculations.
12 A separate Worksheet is provided for use in answering Question 29.

SECTION A

For questions 1 to 20 in this section of the paper, an answer is recorded on the answer sheet by indicating the choice A, B, C, D or E by a stroke made in ink in the appropriate box of the answer sheet—see the example below.

EXAMPLE

The energy unit measured by the electricity meter in your home is the

  A ampere

  B kilowatt-hour

  C watt

  D coulomb

  E volt.

The correct answer to the question is B—kilowatt-hour. Record your answer by drawing a heavy vertical line joining the two dots in the appropriate box on your answer sheet in the column of boxes headed B. The entry on your answer sheet would now look like this:

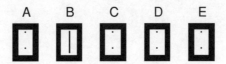

If after you have recorded your answer you decide that you have made an error and wish to make a change, you should cancel the original answer and put a vertical stroke in the box you now consider to be correct. Thus, if you want to change an answer D to an answer B, your answer sheet would look like this:

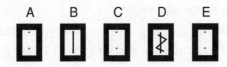

If you want to change back to an answer which has already been scored out, you should enter a tick (✓) to the RIGHT of the box of your choice, thus:

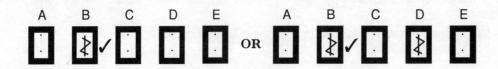

SECTION A

Answer questions 1–20 on the answer sheet.

1. A car accelerates from 4·0 m/s to 20 m/s in 5·0 s. The acceleration of the car is

   A $0·5 \text{ m/s}^2$
   B $3·2 \text{ m/s}^2$
   C $4·0 \text{ m/s}^2$
   D $4·8 \text{ m/s}^2$
   E 16 m/s^2.

2. An athlete runs 30 m East and then 40 m West.

   Which row correctly shows the distance gone and the displacement from the starting point?

	Distance	Displacement
A	10 m	10 m East
B	10 m	10 m West
C	10 m	70 m East
D	70 m	10 m West
E	70 m	10 m East

3. Two forces act on a block of mass 2 kg as shown.

   The block is initially at rest.

   The speed-time graph for the block is

   A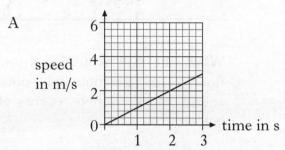

   B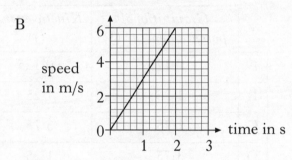

   C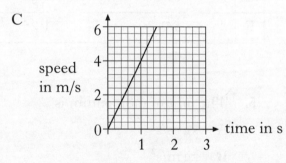

   D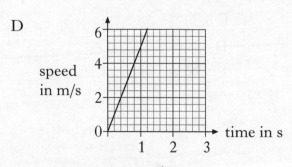

   E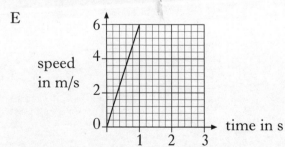

4. A ball of mass 0·50 kg is released from a height of 1·00 m.

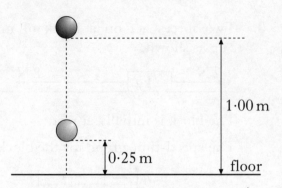

When the ball is 0·25 m from the floor, the gravitational potential energy and the kinetic energy of the ball are

	Gravitational potential energy (J)	Kinetic energy (J)
A	0·125	0·125
B	1·25	1·25
C	1·25	3·75
D	3·75	1·25
E	5·00	1·25

5. The unit of momentum is

A   kg m/s

B   kg m/s^2

C   J

D   N m

E   N/kg.

6. A vehicle of mass 600 kg travelling at 12 m/s hits a stationary vehicle of mass 1200 kg.

The vehicles lock together.

The velocity of the vehicles immediately after the collision is

A   3·0 m/s

B   4·0 m/s

C   6·0 m/s

D   8·0 m/s

E   12 m/s.

7. A solid is placed in an insulated flask and heated continuously with an immersion heater. The sketch graph below shows how the temperature of the contents of the flask changes with time.

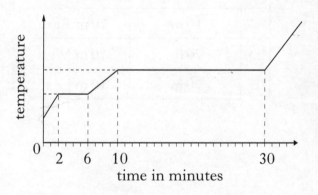

After 5 minutes the contents of the flask are

A   in the solid state

B   in the liquid state

C   a mixture of solid and liquid

D   in the gaseous state

E   a mixture of liquid and gas.

8. Water enters a solar panel at 15 °C and leaves at 20 °C.

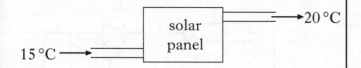

The specific heat capacity of water is 4200 J/kg °C.

4 kg of water passes through the panel every minute.

The heat energy gained by the water in 1 minute is

A 16 800 J
B 84 000 J
C 252 000 J
D 336 000 J
E 1 000 800 J.

9. Which row correctly shows the units of charge, current and power?

	Charge	Current	Power
A	coulomb	ampere	watt
B	coulomb	ampere	joule
C	volt	ampere	watt
D	volt	ampere	joule
E	volt	coulomb	watt

10. A 5 Ω and a 20 Ω resistor are connected in parallel.

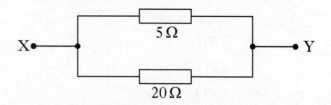

The resistance between X and Y is

A 0·25 Ω
B 4·0 Ω
C 12·5 Ω
D 15 Ω
E 25 Ω.

11. A kettle is rated at 230 V, 2300 W.

The charge passing through the element of the kettle in 200 s is

A 20 C
B 2000 C
C 46 000 C
D 460 000 C
E 529 000 C.

12. A circuit contains a battery, a resistor and four LEDs P, Q, R and S.

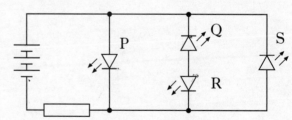

Which LED(s) is/are lit?

A P only
B S only
C P and R only
D Q and S only
E P and S only

13. An electric motor connected to a 12 V supply draws a current of 0·5 A. The energy supplied to the motor in 30 s is

    A   6 J
    B   15 J
    C   180 J
    D   360 J
    E   720 J.

14. Which of the following equations can be used to find the power supplied to a resistor?

    I $P = VI$

    II $P = I^2 R$

    III $P = \dfrac{V^2}{R}$

    A   I only
    B   II only
    C   III only
    D   I and II only
    E   I, II and III

15. A resistor and LED are connected in series across a 5 V d.c. supply.

    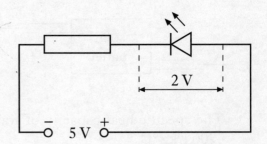

    The current in the LED is 20 mA.
    The voltage across the LED is 2 V.

    The resistance of the resistor is

    A   0·10 Ω
    B   0·15 Ω
    C   100 Ω
    D   150 Ω
    E   250 Ω.

16. A water wave is shown below.

    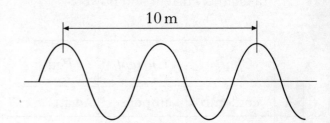

    The speed of the wave is 2·0 m/s.
    The frequency of the wave is

    A   0·2 Hz
    B   0·4 Hz
    C   2·5 Hz
    D   10 Hz
    E   20 Hz.

17. An object is placed 10 cm from a converging lens of focal length 15 cm.

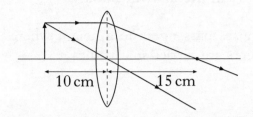

The image formed by the lens is

A inverted and the same size as the object

B inverted and smaller than the object

C inverted and larger than the object

D upright and smaller than the object

E upright and larger than the object.

18. A ray of light strikes a plane mirror at an angle of 40° to the mirror surface.

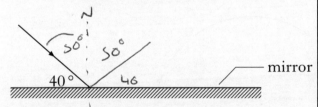

Which row shows the correct values of angle of incidence and angle of reflection for this ray?

	Angle of incidence in degrees	Angle of reflection in degrees
A	40	40
B	40	50
C	40	140
D	50	40
E	50	50

19. A student makes the following three statements.

I Alpha particles produce much greater ionisation density than beta particles or gamma rays.

II Alpha particles are fast moving electrons.

III Alpha particles can be stopped by a piece of paper.

Which of these statements is/are correct?

A I only

B II only

C III only

D I and III only

E I, II and III

20. Measurements are made of the absorbed dose and dose equivalent received by workers in the nuclear industry. The relationship between absorbed dose and dose equivalent is

A $Q = DH$

B $D = HQ$

C $H = DQ$

D $H = \dfrac{Q}{D}$

E $H = \dfrac{D}{Q}$.

[Turn over

SECTION B

Write your answers to questions 21–31 in the answer book.

21. An observation wheel rotates slowly and raises passengers to a height where they can see across a large city. The passengers are carried in capsules.

    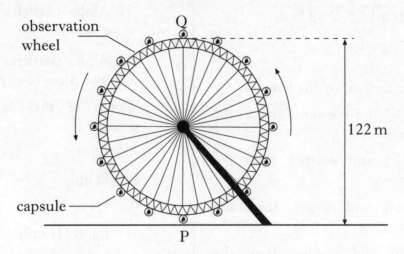

    (a) Each capsule is raised through a height of 122 m as it moves from P to Q.

    Each capsule with passengers has a total mass of 2750 kg.

    Calculate the gravitational potential energy gained by a capsule with passengers. **2**

    (b) The wheel is rotated by a driving force of 200 kN.

    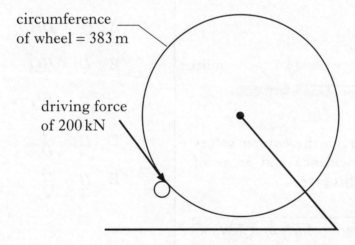

    (i) For one revolution, the driving force is applied through the circumference of the wheel, a distance of 383 m.

    Calculate the work done by the driving force for one revolution. **2**

    (ii) The observation wheel rotates once every 30 minutes.

    Calculate the power delivered to the wheel. **2**

    (c) The driving system does not supply all the gravitational potential energy gained by the upward moving capsules. Explain how these capsules gain the additional energy required. **2**

    (8)

22. Table tennis players can practise using a device which fires balls horizontally.

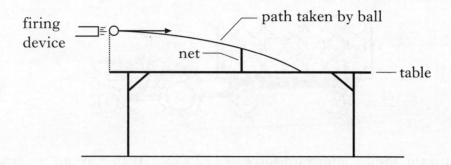

The following graphs describe the horizontal and vertical motions of a ball from the instant it leaves the device until it bounces on the table 0·25 s later.

The effects of air resistance are assumed to be negligible.

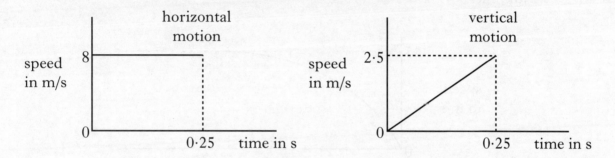

(a) Explain why the shape of the path taken by the ball is curved. 2

(b) (i) What is the instantaneous speed of the ball as it leaves the device? 1

(ii) Describe a method of measuring the instantaneous speed of the ball as it leaves the device. 3

(iii) Calculate the height above the table at which the ball is released. 2

(c) The device is adjusted to fire a second ball which lands at the end of the table.

The height and position of the device are not changed.

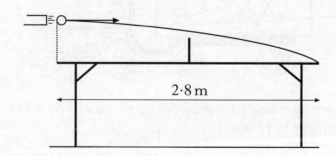

The length of the table is 2·8 m.

Assuming that the effects of air resistance are negligible, calculate the instantaneous speed of the second ball as it leaves the device. 2

(10)

23. A tractor and a loaded trailer have a total mass of 9500 kg.

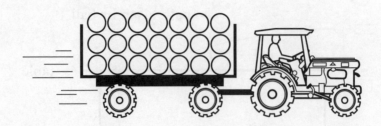

(a) The tractor applies a forward force of 15 250 N. At the instant the tractor and trailer move off the total frictional force is 1000 N.

Calculate the initial acceleration of the tractor and trailer. **3**

(b) The following graph shows how the speed of the tractor and trailer varies with time.

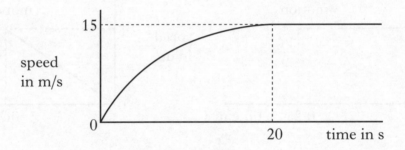

The tractor continues to apply a forward force of 15 250 N. State the size of the frictional force after 20 s. **1**

(c) On a second journey the trailer is loaded in a different way. The total mass of the tractor and trailer is again 9500 kg.

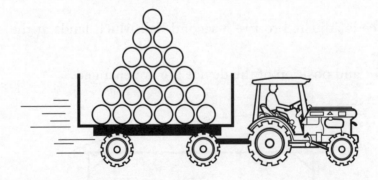

The tractor again applies a forward force of 15 250 N. The maximum speed on this journey is 12 m/s.

Explain, in terms of forces, why the maximum speed on this journey is less than the maximum speed in part (b). **2**

(6)

24. A technician uses the soldering iron shown when connecting electrical components.

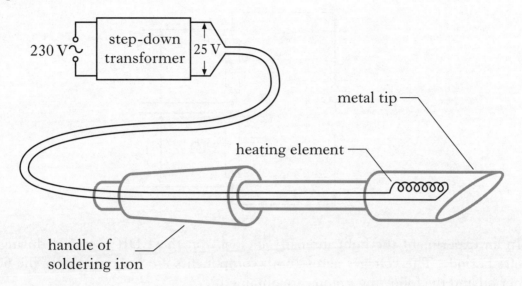

The heating element is used to raise the temperature of the metal tip above the melting point of solder. The heating element is rated at 25 V, 90 W.

(a) A step-down transformer is used to reduce voltage from the mains value of 230 V to 25 V.

There are 1840 turns on the primary coil of the transformer.

Calculate the number of turns on the secondary coil. 2

(b) The heating element is switched on for 50 s.

Calculate the electrical energy supplied to the element. 2

(c) The metal tip is made of copper and has a mass of 0·03 kg. The temperature of the metal tip rises from 20 °C to 370 °C during the period that the element is switched on.

The specific heat capacity of copper is 386 J/kg °C.

Calculate the heat energy gained by the metal tip. 2

(d) Explain why the heat energy gained by the metal tip is less than the electrical energy supplied to the element. 2

(e) A device which uses a thermocouple is used to measure the temperature of the metal tip. State the energy change which takes place in a thermocouple. 1

(9)

[Turn over

25. A light dependent resistor and a 2 kΩ resistor are connected in series across a d.c. supply.

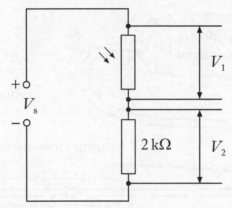

In an experiment the light intensity incident on the LDR is varied during a 60 s period. The voltages across both components are measured over the 60 s period and the following graphs are obtained.

graph 1

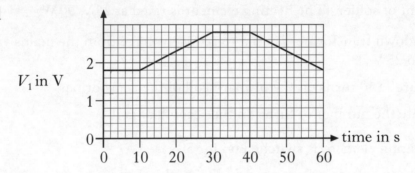

graph 2

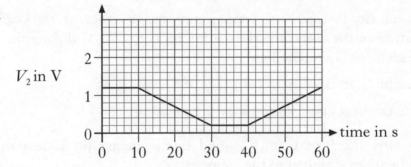

(a) (i) State the values of V_1 and V_2 at the start of the 60 s period. 1

    (ii) State the value of the supply voltage V_s. 1

(b) Calculate the resistance of the light dependent resistor at the start of the 60 s period. 2

(c) Is the light intensity incident on the LDR increasing or decreasing during the time interval between 10 s and 30 s? You must explain your answer. 3

25. (continued)

(d) An additional circuit is connected across the 2 kΩ resistor.

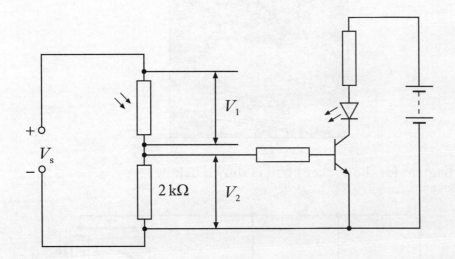

The experiment is now repeated and identical graphs are obtained.

A student observes that the LED is lit between 0 s and 20 s, off between 20 s and 50 s and lit between 50 s and 60 s.

Use graph 2 to explain why the LED is off between 20 s and 50 s.

2

(9)

[Turn over

26. A cooker hood contains two 40 W lamps and an extractor fan.

A circuit diagram for the cooker hood is shown below.

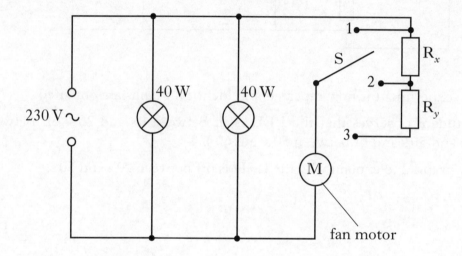

(a) Calculate the current drawn by one lamp. 2

(b) Calculate the resistance of one lamp. 2

(c) The speed of the fan motor is varied by moving switch S to position 1, 2 or 3.

State the position of S for maximum speed of the extractor fan motor. You must explain your reason for selecting that position. 2

(d) When S is in position 2 the voltage across the motor is 180 V and the current through the motor is 0·25 A.

Calculate the resistance of R_x. 3

(9)

27. At a concert a musician sings into a microphone connected to an amplifier.

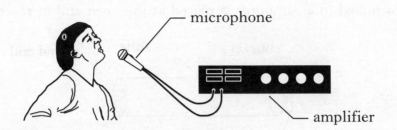

The sound energy causes a diaphragm in the microphone to move up and down. A coil of wire is attached to the diaphragm so that it moves up and down with the diaphragm.

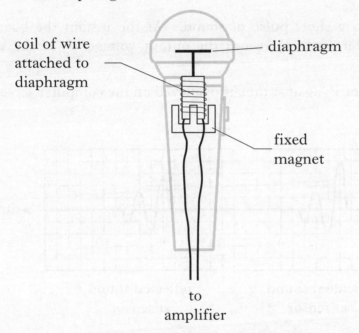

(a) Explain how an a.c. voltage is induced in the coil. 2

(b) State **two** changes in the design of the microphone which would result in a greater induced voltage. 2

(c) A voltage of 2 mV from the microphone is applied to the input terminals of the amplifier. The output voltage of the amplifier is 0·5 V.

Calculate the voltage gain of the amplifier. 2

(6)

[Turn over

28. A buzzer is placed in front of the open end of a tube. The tube is closed at the other end.

A sound sensor linked to a computer is placed at the open end of the tube as shown.

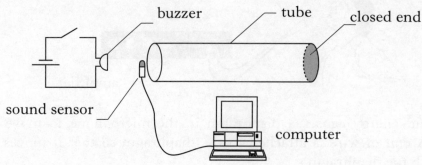

The buzzer produces a short pulse of sound. At the instant the buzzer is operated the computer starts to record the output voltage V_o of the sound sensor.

The following graph of V_o against time is displayed on the computer screen.

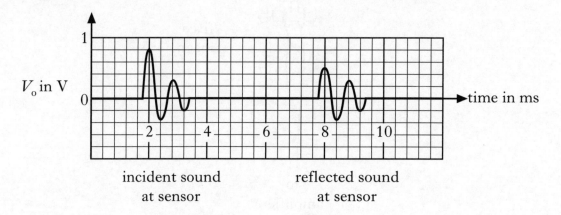

(a) Explain why the amplitude of the reflected sound is less than the amplitude of the incident sound. **1**

(b) State the time between the first peak of the incident sound and the first peak of the reflected sound arriving at the sound sensor. **1**

(c) The speed of sound in air is 340 m/s.

Calculate the length of the tube. **3**

(d) The frequency of the pulse is 1250 Hz.

Calculate the wavelength of the pulse. **2**

(7)

Marks

29. Rays of light enter glass prisms as shown in diagrams 1 and 2.

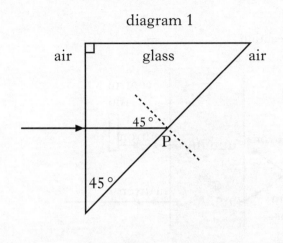
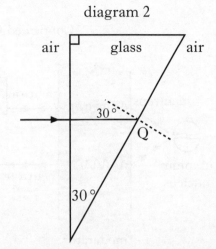

The critical angle for glass is 42°.

(a) Using **Worksheet Q29**, complete diagram 1 to show the path of the ray after it strikes point P. 2

(b) Using **Worksheet Q29**, complete diagram 2 to show the path of the ray after it strikes point Q. 2

(4)

30. The following table contains information about two radioactive sources used in medicine.

Radioactive source	*Activity* (MBq)	*Half-life* (days)
R	1600	8
S	80	74

(a) Calculate the number of decays of source R in 30 s. 2

(b) These radioactive sources can be disposed of after their activity has fallen below 40 MBq.

Show, by calculation, which source, R or S, will be the first to reach an activity of 40 MBq. 3

(c) State **two** safety precautions which should be taken when handling radioactive sources. 2

(7)

[Turn over for Question 31 on *Page eighteen*

31. A simplified model of a controlled chain reaction in a nuclear reactor is shown below.

Controlled Chain Reaction

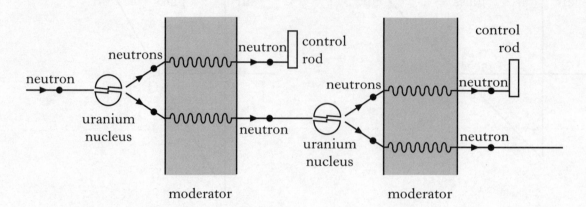

(a) (i) Name the type of nuclear reaction that takes place in the reactor. **1**

(ii) State the purpose of the moderator. **1**

(iii) How could the chain reaction process be stopped? **1**

(b) State **one advantage** and **one disadvantage** of using nuclear power for the generation of electricity. **2**

(5)

[END OF QUESTION PAPER]

X069/202

| NATIONAL QUALIFICATIONS 2002 | WEDNESDAY, 22 MAY 1.00 PM – 3.00 PM | PHYSICS INTERMEDIATE 2 Worksheet Q29 |

Fill in these boxes and read what is printed below.

Full name of centre

Town

Forename(s)

Surname

Date of birth
Day Month Year

Scottish candidate number

Number of seat

To be inserted inside the front cover of the candidate's answer book and returned with it.

WORKSHEET Q29

(a)

diagram 1

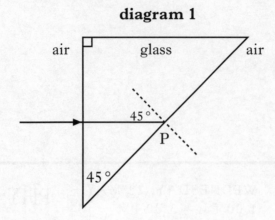

(b)

diagram 2

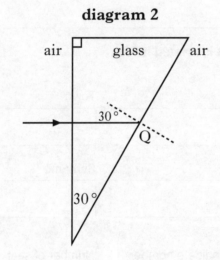

[END OF WORKSHEET]

2003 | Intermediate 2

[BLANK PAGE]

X069/201

NATIONAL
QUALIFICATIONS
2003

MONDAY, 19 MAY
1.00 PM – 3.00 PM

PHYSICS
INTERMEDIATE 2

Read Carefully

1 All questions should be attempted.

Section A (questions 1 to 20)

2 Check that the answer sheet is for Physics Intermediate 2 (Section A).
3 Answer the questions numbered 1 to 20 on the answer sheet provided.
4 Fill in the details required on the answer sheet.
5 Rough working, if required, should be done only on this question paper, or on the first two pages of the answer book provided—**not** on the answer sheet.
6 For each of the questions 1 to 20 there is only **one** correct answer and each is worth 1 mark.
7 Instructions as to how to record your answers to questions 1–20 are given on page two.

Section B (questions 21 to 30)

8 Answer the questions numbered 21 to 30 in the answer book provided.
9 Fill in the details on the front of the answer book.
10 Enter the question number clearly in the margin of the answer book beside each of your answers to questions 21 to 30.
11 Care should be taken to give an appropriate number of significant figures in the final answers to calculations.

SECTION A

For questions 1 to 20 in this section of the paper, an answer is recorded on the answer sheet by indicating the choice A, B, C, D or E by a stroke made in ink in the appropriate box of the answer sheet—see the example below.

EXAMPLE

The energy unit measured by the electricity meter in your home is the

    A ampere

    B kilowatt-hour

    C watt

    D coulomb

    E volt.

The correct answer to the question is B—kilowatt-hour. Record your answer by drawing a heavy vertical line joining the two dots in the appropriate box on your answer sheet in the column of boxes headed B. The entry on your answer sheet would now look like this:

If after you have recorded your answer you decide that you have made an error and wish to make a change, you should cancel the original answer and put a vertical stroke in the box you now consider to be correct. Thus, if you want to change an answer D to an answer B, your answer sheet would look like this:

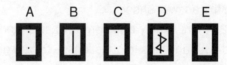

If you want to change back to an answer which has already been scored out, you should enter a tick (✓) to the RIGHT of the box of your choice, thus:

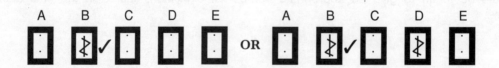

SECTION A

Answer questions 1–20 on the answer sheet.

1. Which row in the table contains only vector quantities?

A	momentum	displacement	force
B	distance	velocity	energy
C	momentum	speed	force
D	force	energy	distance
E	speed	distance	time

2. The diagram represents two forces acting on an object.

   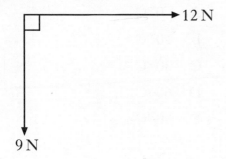

   The magnitude of the resultant force is

   A 3 N
   B 10 N
   C 11 N
   D 15 N
   E 21 N.

3. A man of mass 80 kg dives from a diving board which is 10 m above water. Neglecting air friction, the kinetic energy of the diver immediately before he hits the water is

   A 14 J
   B 800 J
   C 1200 J
   D 4000 J
   E 8000 J.

4. A block of wood of mass 2 kg is pulled along a bench by a horizontal force of 6 N.

   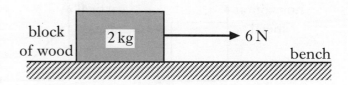

   A constant frictional force of 2 N acts on the block.

   The acceleration of the block is

   A 0.25 m/s^2
   B 0.5 m/s^2
   C 2 m/s^2
   D 3 m/s^2
   E 4 m/s^2.

   [Turn over

5. A ball is kicked horizontally off the edge of a cliff and lands in the sea. Which pair of graphs shows the horizontal and vertical speeds of the ball during its flight? The effect of air friction should be ignored.

A
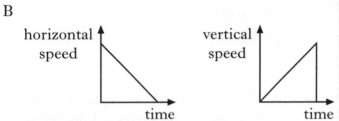

B
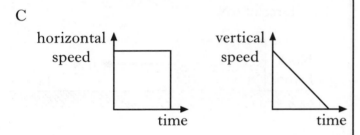

C
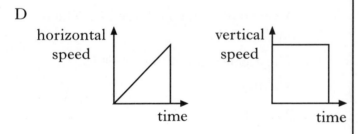

D
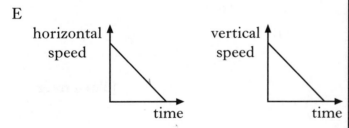

E
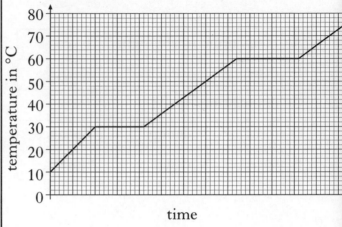

6. A block of wax, initially in the solid state, is heated. The graph below shows how the temperature of the wax changes with time.

At what temperature does the wax melt?

A 0 °C
B 10 °C
C 30 °C
D 60 °C
E 80 °C

7. A motor has an efficiency of 60%. The input energy to this motor is 200 J. The output energy of the motor is

A $\dfrac{200}{100 \times 60}$ J

B $\dfrac{200}{60}$ J

C $\dfrac{200 \times 60}{100}$ J

D $\dfrac{200 \times 100}{60}$ J

E 60×200 J.

8. A room in a house has two lamps X and Y. With different switch positions, either lamp X or lamp Y or both lamps X and Y can be lit.

Which circuit allows the lamps to operate in this way?

A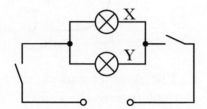

B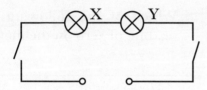

C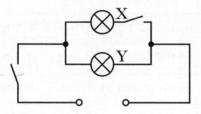

D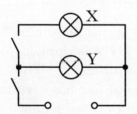

E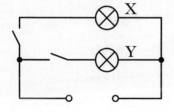

9. The information shown applies to an electric iron.

ELECTRIC IRON	
Operating voltage	230 V
Power	2·3 kW
Resistance	23 Ω

The iron is switched on for 10 minutes. How much electrical energy is converted to heat energy during this time?

A 5290 J

B 529 000 J

C 717 600 J

D 1 380 000 J

E 2 116 000 J

10. The frequency of the mains supply is

A 0·02 Hz

B 5 Hz

C 50 Hz

D 230 Hz

E 240 Hz.

[Turn over

11. Two resistors are connected in series with a 12 volt d.c. supply.

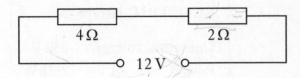

The current in the 2 Ω resistor is 2 A. Which row of the table gives the current in the 4 Ω resistor and the voltage across the 4 Ω resistor?

	Current in A	Voltage in V
A	1	4
B	1	12
C	2	8
D	2	12
E	4	8

12. A lap top computer is connected to the output of a transformer as shown.

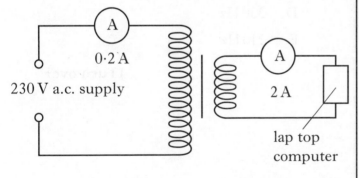

The transformer is 100% efficient. The resistance of the lap top computer is

A 4·6 Ω

B 11·5 Ω

C 46 Ω

D 115 Ω

E 1150 Ω.

13. A magnet is placed inside a coil of wire connected to a voltmeter as shown below.

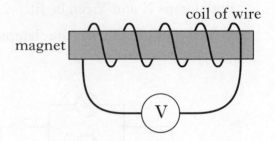

The magnet can be moved either to the left or to the right.

The coil of wire can also be moved to the left or to the right.

Which of the following produces a reading of zero on the meter?

	Movement of magnet	Movement of coil
A	to the right at 0·2 m/s	to the right at 0·2 m/s
B	to the left at 0·2 m/s	to the right at 0·2 m/s
C	to the left at 0·2 m/s	stationary
D	to the right at 0·2 m/s	to the left at 0·2 m/s
E	to the right at 0·2 m/s	to the right at 0·4 m/s

14. A signal of voltage 2 V and frequency 100 Hz is applied to the input of an amplifier. The output of the amplifier has a voltage of 10 V. The output frequency of the amplifier is

A 5 Hz

B 20 Hz

C 50 Hz

D 100 Hz

E 500 Hz.

15. Which row in the table correctly shows input and output devices?

	Input device	Output devices	
A	microphone	loudspeaker	LED
B	solar cell	thermocouple	LED
C	loudspeaker	microphone	relay
D	LED	loudspeaker	solar cell
E	thermocouple	microphone	LED

16. A student places an object 250 mm from a converging lens of focal length of 100 mm.

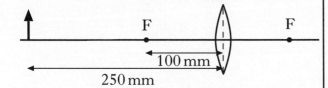

The image formed by the lens is

A inverted and the same size as the object

B inverted and smaller than the object

C inverted and larger than the object

D upright and smaller than the object

E upright and larger than the object.

17. In a water tank, 10 waves pass a point in 2 seconds. The speed of the waves is 0·4 m/s. The wavelength of the waves is

A 0·005 m

B 0·02 m

C 0·04 m

D 0·08 m

E 2 m.

18. Sound is a longitudinal wave. When a sound wave travels through air the particles of air

A move continuously away from the source

B move continuously towards the source

C vibrate at random

D vibrate at 90° to the wave direction

E vibrate along the wave direction.

[Turn over

19. Which sign is used to indicate the presence of radioactive material?

A

B

C

D

E

20. A student writes the following statements.

 I Alpha radiation is part of the electromagnetic spectrum.

 II Alpha radiation is more ionising than beta or gamma radiation.

 III Alpha radiation is more penetrating than beta or gamma radiation.

Which of the statements is/are true?

A I only
B II only
C III only
D I and II only
E I and III only

SECTION B

Write your answers to questions 21–30 in the answer book.

21. A theme park has a water splash ride. A carriage loaded with passengers is raised through a height of 30 m to the top of the ride. The combined mass of the carriage and the passengers is 1400 kg.

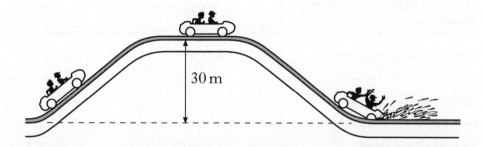

(a) Calculate the gain in gravitational potential energy of the carriage and passengers when it is taken to the top of the ride. **2**

(b) The carriage and passengers stop briefly before being released at the top of the ride. A speed-time graph of the motion of the carriage from the top of the ride is shown below.

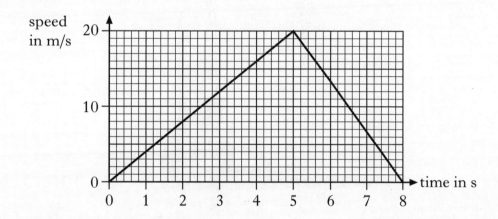

(i) Calculate the acceleration of the carriage from the top of the ride to the point where it reaches the water. **2**

(ii) Calculate the distance travelled by the carriage from the top of the ride to the point where it comes to rest. **2**

(iii) A test run is carried out without any water in the ride. The carriage travels a longer distance before it comes to rest. Explain why this happens. **1**

(7)

[Turn over

Marks

22. A spacecraft travels through space between planet X and planet Y. Information on these planets is shown in the table below.

	planet X	planet Y
Gravitational field strength on surface	8·4 N/kg	13 N/kg
Surface temperature	17·0 °C	9·0 °C
Atmosphere	No	Yes
Period of rotation	48 hours	17 hours

The spacecraft has a total mass of $2·5 \times 10^6$ kg.

The spacecraft engines produce a total force of $3·8 \times 10^7$ N.

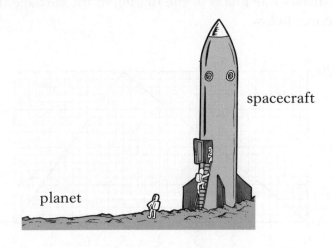

(a) The spacecraft is initially on planet X.

   (i) Calculate the weight of the spacecraft when it is on the surface of planet X. **2**

   (ii) Sketch a diagram showing the forces acting on the spacecraft just as it lifts off from planet X. You must name these forces and show their directions. **2**

   (iii) Calculate the acceleration of the spacecraft as it lifts off from planet X. **3**

(b) On another occasion, the spacecraft lifts off from planet Y. The mass and engine force of the spacecraft are the same as before. Is the acceleration as it lifts off from planet Y less than, more than or equal to the acceleration as it lifts off from planet X?

You **must** give a reason for your answer using information contained in the table above. **2**

(9)

Marks

23. A student investigates collisions using model cars A and B.

Car B is fitted with a piece of card and the edge of the card is placed close to a light gate attached to a timer as shown.

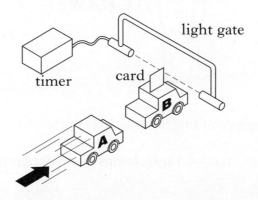

(a) In one experiment car A is moving directly towards car B which is stationary. The cars collide and stick together. After the collision the card passes through the light gate.

The student records the following measurements.

Mass of car A = 1·6 kg
Mass of car B = 1·0 kg
Speed of car B before collision = 0 m/s
Length of card = 100 mm
Time on timer = 0·05 s

(i) Calculate the speed of the cars **after** the collision. 2

(ii) Use your answer for part (i), and information contained in the student's measurements, to calculate the speed of car A immediately before the collision. 2

(b) In a second experiment car A is moving with a different speed directly towards stationary car B. The cars again collide and stick together. The cars have a speed of 4 m/s after the collision.

(i) Calculate the total kinetic energy of the cars after the collision. 2

(ii) After this collision the cars move in a straight line and come to rest. The frictional force acting on the cars is 2·6 N. Calculate the distance travelled by the cars after the collision. 2

(c) In each experiment the edge of card is placed close to the light gate before the collision. Explain why. 1

(9)

[*Turn over*

24. One type of lamp used for Christmas tree sets is rated as follows.

(a) Show that the resistance of one lamp is 46 Ω. **2**

(b) In one arrangement, ten of these lamps are connected in series to the mains as shown.

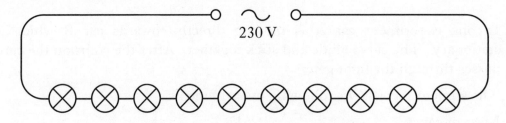

   (i) Show that the voltage across each lamp is 23 V. **1**

   (ii) State **one** disadvantage of wiring lamps in this way. **1**

(c) In another arrangement, the ten lamps are connected in parallel as shown.

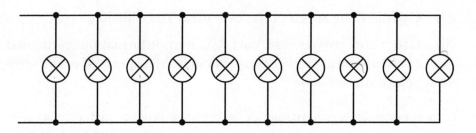

   (i) Calculate the total resistance of this arrangement of lamps. **2**

   (ii) This arrangement of lamps cannot be connected directly to the mains, but it can be connected to the mains by using a transformer.

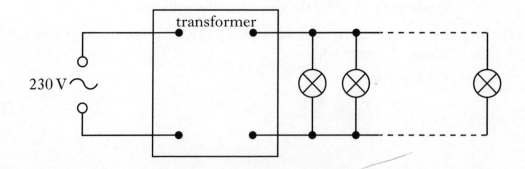

24. (c) (ii) (continued)

A technician investigates the relationship between the primary voltage and secondary voltage of the transformer. The following graph is obtained from the technician's results.

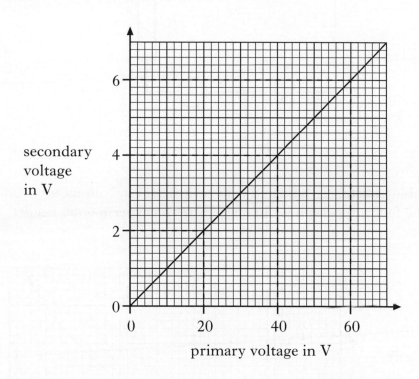

(A) Calculate the output voltage from the transformer when it is attached to the mains supply of 230 V.   **2**

(B) The parallel arrangement is connected to the mains through this transformer. **Explain** whether the lamps operate at normal brightness.   **1**

(9)

[Turn over

25. An LED is connected in the circuit shown.

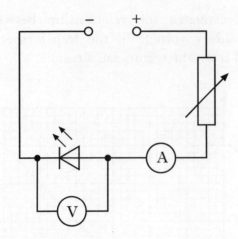

The variable resistor is adjusted and voltmeter and ammeter readings are taken. The following graph is obtained from the experimental results.

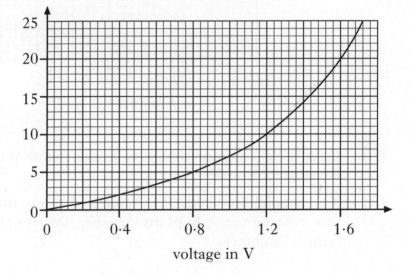

(a) Using information from the graph, determine how the resistance of the LED changes as the voltage across it is increased.

You **must** justify your answer by calculation.   3

25. (continued)

(b) The LED is now connected into a circuit with a resistor R as shown.

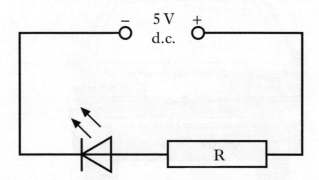

(i) The current in the LED is 20 mA. Using the graph on *Page fourteen*, state the voltage across the LED. **1**

(ii) Calculate the resistance of resistor R. **3**

(7)

[Turn over

26. (a) A heater is used to melt ice on the rear window of a car.

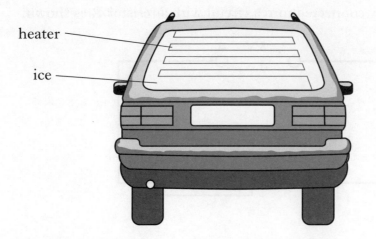

(i) Calculate the heat energy required to melt 0·05 kg of ice.
(Latent heat of fusion of ice = $3·34 \times 10^5$ J/kg) **2**

(ii) The heater takes 5 minutes to melt 0·05 kg of ice. Assuming all the energy is used to melt the ice, calculate the output power of the heater. **2**

(b) The car has a warning light which comes on when the outside temperature falls below 3 °C. The circuit for the warning light is shown.

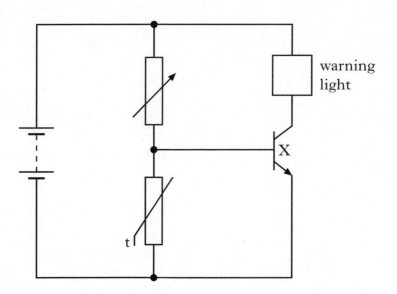

(i) Name component X. **1**

(ii) What happens to the resistance of the thermistor as the temperature falls? **1**

(iii) Explain how the circuit operates so that the warning light comes on when the temperature falls below 3 °C. **2**

(8)

27. A student reads the following article about nuclear power.

"In a nuclear reactor, uranium nuclei in fuel rods are bombarded with neutrons. A uranium nucleus may absorb a neutron and then break up into two smaller nuclei releasing further neutrons and energy."

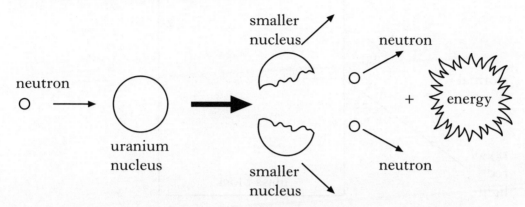

(a) (i) A **nucleus** contains 2 types of particle. Name these particles.   1

(ii) What is the name given to the process shown in the diagram?   1

(iii) Explain why fuel rods have to be replaced after a certain time.   1

(iv) Explain why the fuel rods that are removed from the reactor are a safety hazard.   1

(b) In a nuclear reactor, 166 MJ of energy is transferred to 2000 kg of coolant. All of this energy is absorbed by the coolant which has a specific heat capacity of 830 J/kg °C. Assuming the coolant does not change state, calculate the rise in temperature of the coolant.   2

(6)

[Turn over

28. (a) A ray of red light is incident on a glass block as shown below.

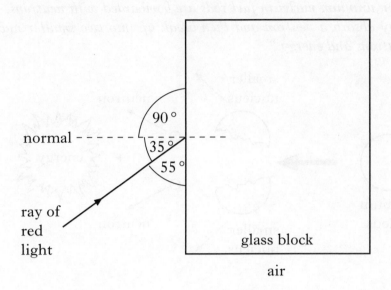

(i) State the size of the angle of incidence.   **1**

(ii) Copy the diagram and complete it to show the path of the ray inside the glass block.   **1**

(b) In another experiment, rays of red light are incident on three semi-circular blocks of glass as shown. Each block is made of a different type of glass.

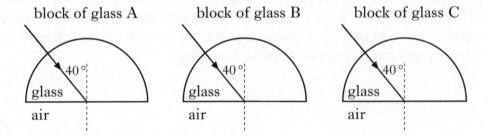

The critical angle for each block of glass is given below.

Glass type	Critical angle
A	38°
B	42°
C	44°

From which block(s) does a ray of light refract through the straight edge? Explain your answer.   **2**

28. (continued)

(c) One particular **short sighted** person requires a lens with a focal length of 200 mm. Six lenses, each of different type and power, are available.

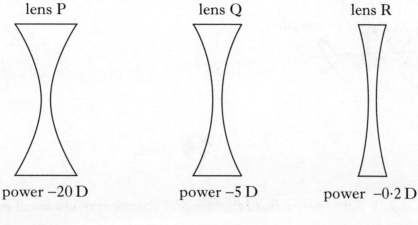

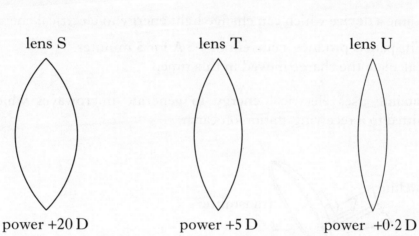

(i) Name the type of lens used to correct for short sight. 1

(ii) From the lenses shown, choose the one that corrects for the short sight of the above person.

You must justify your answer by calculation. 2

(7)

[Turn over

29. A satellite orbiting the earth has large panels as shown.

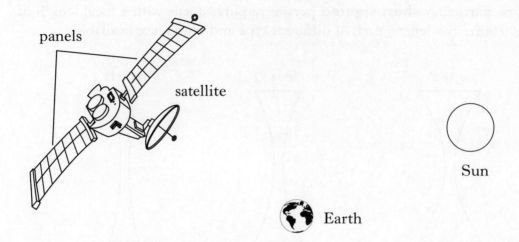

The panels absorb light energy from the sun and change it to electrical energy.

(a) (i) Name a device which can change light energy to electrical energy. **1**

(ii) The panels produce a current of 4·5 A for 5 minutes. Calculate the charge moved in this time. **2**

(b) The satellite uses electrical energy to generate microwaves which are transmitted to a receiving station on earth.

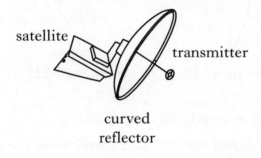

Explain how the curved reflector on the satellite aids the **transmission** of the microwaves. You must sketch a diagram as part of your answer. **2**

29. (continued)

(c) Microwaves are part of the electromagnetic spectrum.

The diagram below shows the electromagnetic spectrum arranged in order of wavelength. Two parts of the spectrum, P and Q, have been omitted. Name the radiations P and Q.

increasing wavelength →

| gamma rays | P | ultra violet | visible | Q | micro waves | radio + TV |

1

(d) All radiations in the electromagnetic spectrum travel at a speed of 3×10^8 m/s in space.

The satellite transmits microwaves on the following three frequencies.

1.0×10^{10} Hz
9.0×10^9 Hz
8.0×10^9 Hz

Calculate the wavelength of the microwaves with the **longest** wavelength.

3

(9)

[Turn over for Question 30 on *Page twenty-two*

Marks

30. Companies delivering radioactive sources have to follow strict safety rules. One rule is that sources must be labelled. The following information is displayed on a label on a radioactive source.

> **RADIOACTIVE SOURCE**
>
> Source: beta and gamma emitter
> Year of delivery: 2003
> Half life: 10 years
> Activity: 20 000 Bq

(a) (i) What is meant by the activity of a source? **1**

(ii) Calculate the activity of the source in year 2043. **2**

(b) After delivery, the source is placed in a thick walled aluminium storage box. Which type of radiation from the source, if either, could penetrate the storage box? You must explain your answer. **2**

(c) A technician handling an **alpha-emitting** source estimates that his hand receives an absorbed dose of 5×10^{-5} Gy. The mass of the technician's hand is 500 g.

(i) Calculate the total energy absorbed by the technician's hand. **2**

(ii) Using information from the table below, calculate the dose equivalent received by his hand.

Type of radiation	Quality factor
Alpha	20
Beta	1
Gamma	1
X rays	1
Slow neutrons	2·3

2

(9)

[END OF QUESTION PAPER]

2004 | Intermediate 2

[BLANK PAGE]

X069/201

NATIONAL
QUALIFICATIONS
2004

FRIDAY, 28 MAY
1.00 PM – 3.00 PM

PHYSICS
INTERMEDIATE 2

Read Carefully

1 All questions should be attempted.

Section A (questions 1 to 20)

2 Check that the answer sheet is for Physics Intermediate 2 (Section A).

3 Answer the questions numbered 1 to 20 on the answer sheet provided.

4 Fill in the details required on the answer sheet.

5 Rough working, if required, should be done only on this question paper, or on the first two pages of the answer book provided—**not** on the answer sheet.

6 For each of the questions 1 to 20 there is only **one** correct answer and each is worth 1 mark.

7 Instructions as to how to record your answers to questions 1–20 are given on page two.

Section B (questions 21 to 31)

8 Answer the questions numbered 21 to 31 in the answer book provided.

9 Fill in the details on the front of the answer book.

10 Enter the question number clearly in the margin of the answer book beside each of your answers to questions 21 to 31.

11 Care should be taken to give an appropriate number of significant figures in the final answers to calculations.

SECTION A

For questions 1 to 20 in this section of the paper, an answer is recorded on the answer sheet by indicating the choice A, B, C, D or E by a stroke made in ink in the appropriate box of the answer sheet—see the example below.

EXAMPLE

The energy unit measured by the electricity meter in your home is the

    A ampere

    B kilowatt-hour

    C watt

    D coulomb

    E volt.

The correct answer to the question is B—kilowatt-hour. Record your answer by drawing a heavy vertical line joining the two dots in the appropriate box on your answer sheet in the column of boxes headed B. The entry on your answer sheet would now look like this:

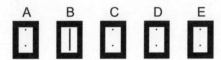

If after you have recorded your answer you decide that you have made an error and wish to make a change, you should cancel the original answer and put a vertical stroke in the box you now consider to be correct. Thus, if you want to change an answer D to an answer B, your answer sheet would look like this:

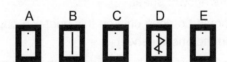

If you want to change back to an answer which has already been scored out, you should enter a tick (✓) to the RIGHT of the box of your choice, thus:

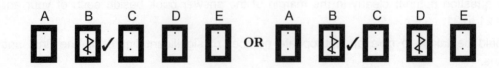

SECTION A

Answer questions 1–20 on the answer sheet.

1. Which of the following shows two physical quantities that have the same unit?

   A Potential energy and work done
   B Momentum and kinetic energy
   C Potential energy and momentum
   D Force and work done
   E Force and mass

2. Near the Earth a mass of 4 kg is falling with a constant velocity.

   The air resistance force and the unbalanced force acting on the mass are

	Air resistance force	Unbalanced force
A	10 N upwards	10 N downwards
B	10 N downwards	50 N downwards
C	40 N upwards	0 N
D	40 N upwards	40 N downwards
E	10 N upwards	0 N

3. The table gives information about the velocities of three objects **X**, **Y** and **Z** for a time interval of 3 seconds. Each object is moving in a straight line.

Time (s)	0	1	2	3
Velocity of X (m/s)	2	4	6	8
Velocity of Y (m/s)	0	1	2	3
Velocity of Z (m/s)	0	2	5	9

   Which of the following statements is/are correct?

   I X moves with constant velocity.
   II Y moves with constant acceleration.
   III Z moves with constant acceleration.

   A I only
   B II only
   C I and II only
   D I and III only
   E II and III only

4. A mass of 1 kg is pulled along a level bench by a horizontal force of 10 N. The acceleration of the mass is 4 m/s^2. The frictional force opposing the motion is

   A 0·25 N
   B 0·40 N
   C 2·5 N
   D 4 N
   E 6 N.

[Turn over

5. A ball is released from point **Q** on a curved rail, leaves the rail horizontally at **R** and lands 1 s later.

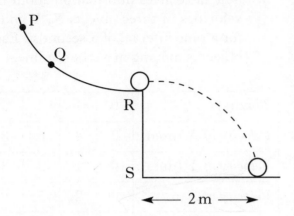

The ball is now released from point **P**.

Which row describes the motion of the ball after leaving the rail?

	Time to land after leaving rail	Distance from S to landing point
A	1 s	less than 2 m
B	less than 1 s	more than 2 m
C	1 s	more than 2 m
D	less than 1 s	2 m
E	more than 1 s	more than 2 m

6. A student makes the following three statements.

   I Momentum is lost in all collisions.

   II Momentum is mass times velocity.

   III Momentum is a vector quantity.

   Which of these statements is/are correct?

   A I only

   B I and II only

   C I and III only

   D II and III only

   E I, II and III

7. Information about water is shown below.

   Specific latent heat of fusion
   $= 3.34 \times 10^5 \, \text{J/kg}$

   Specific heat capacity
   $= 4.18 \times 10^3 \, \text{J/kg}°\text{C}$

   Specific latent heat of vaporisation
   $= 2.26 \times 10^6 \, \text{J/kg}$

   The heat energy required to turn 0·25 kg of water at 100 °C into steam at 100 °C is

   A $0.25 \times 3.34 \times 10^5$ J

   B $0.25 \times 4.18 \times 10^3$ J

   C $4.18 \times 10^3 \times 0.25 \times 100$ J

   D $2.26 \times 10^6 \times 0.25 \times 100$ J

   E $0.25 \times 2.26 \times 10^6$ J.

8. Three resistors are connected as shown.

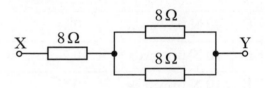

   The total resistance between X and Y is

   A 4 Ω

   B 8 Ω

   C 12 Ω

   D 16 Ω

   E 24 Ω.

9. In the circuit shown, switch S is initially open.

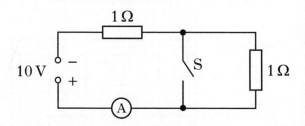

Switch S is now closed.

Which row in the table shows the current with S open and the current with S closed?

	Current with S open	Current with S closed
A	0·1 A	0·2 A
B	0·2 A	0·1 A
C	5 A	2·5 A
D	5 A	10 A
E	10 A	5 A

10. Three identical resistors are connected with three ammeters to a d.c. supply as shown.

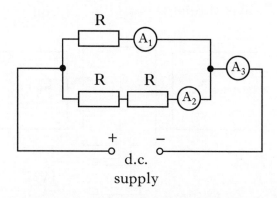

The reading on A_3 is 0·6 A.

Which row shows the readings on A_1 and A_2?

	Ammeter A_1	Ammeter A_2
A	0·2 A	0·4 A
B	0·3 A	0·3 A
C	0·4 A	0·2 A
D	0·6 A	0·3 A
E	0·6 A	0·6 A

11. Consider the following circuit.

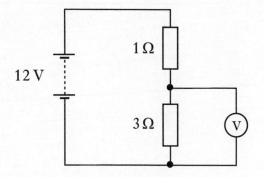

The reading on the voltmeter is

A 3 V
B 4 V
C 8 V
D 9 V
E 12 V.

12. A voltage is induced in a coil when it is rotated in a magnetic field.

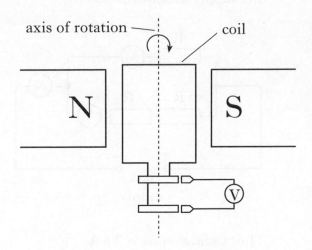

Which of the following combinations of changes produces the greatest increase in the induced voltage?

	Strength of magnetic field	Number of turns in the coil	Speed of rotation of the coil
A	decrease	increase	increase
B	increase	decrease	increase
C	increase	increase	decrease
D	decrease	decrease	decrease
E	increase	increase	increase

13. The graph shows the relationship between the voltage across a resistor and the current in the resistor.

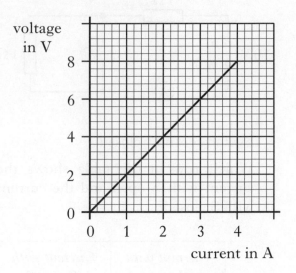

The resistance of the resistor is

A $0.5\,\Omega$

B $2\,\Omega$

C $4\,\Omega$

D $12\,\Omega$

E $32\,\Omega$.

14. Identical thermistors T_1 and T_2 are connected with lamps L_1 and L_2 as shown.

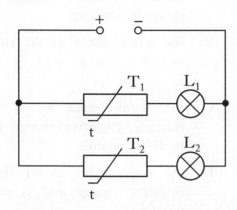

The resistance of these thermistors falls when their temperature rises.

T_1 is heated, T_2 is not heated.

What happens to the brightness of the lamps?

	Brightness of L_1	Brightness of L_2
A	gets dimmer	stays the same
B	stays the same	stays the same
C	gets brighter	gets brighter
D	gets dimmer	gets brighter
E	gets brighter	stays the same

15. The diagram shows part of the electromagnetic spectrum.

P	Visible light	Q	X-rays

The radiations in regions P and Q are

	Region P	Region Q
A	infrared	ultraviolet
B	ultraviolet	microwaves
C	ultraviolet	infrared
D	infrared	microwaves
E	microwaves	ultraviolet

16. Which of the following is a longitudinal wave?

A Water wave

B Radio wave

C Gamma ray

D Sound wave

E Light wave

[Turn over

17. The diagram shows a ray of light incident on the centre of the straight edge of a semicircular glass block.

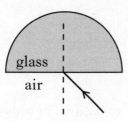

Which diagram shows the path of the ray through the block?

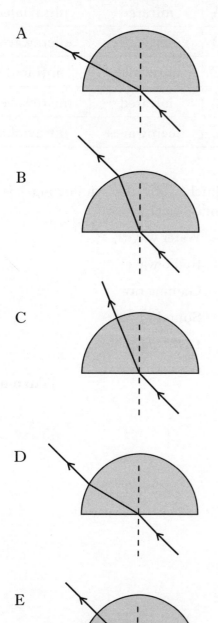

18. An alpha particle has

A the same mass as a helium nucleus, positive charge and is strongly ionising

B the same mass as an electron, negative charge and is weakly ionising

C the same mass as a helium nucleus, negative charge and is weakly ionising

D the same mass as an electron, negative charge and is strongly ionising

E the same mass as a helium nucleus, positive charge and is weakly ionising.

19. A patient's thyroid gland is exposed to radiation. Information about the radiation and the dose received by the gland is shown.

   Absorbed dose   = 500 µGy
   Energy absorbed = 15 µJ
   Quality factor   = 20

The mass of the thyroid gland is

A 0·01 kg
B 0·03 kg
C 0·04 kg
D 0·33 kg
E 0·75 kg.

20. The activity of a sample of a radioactive substance is 80 Bq. The half-life of the substance is 4 hours.

The time for the activity to fall to 10 Bq is

A 4 hours
B 6 hours
C 8 hours
D 12 hours
E 20 hours.

SECTION B

Write your answers to questions 21–31 in the answer book.

21. A cart A of mass 1·2 kg is held at point P on a slope. P is 0·20 m above a horizontal surface. A second cart B of mass 2·8 kg is placed close to the bottom of the slope as shown.

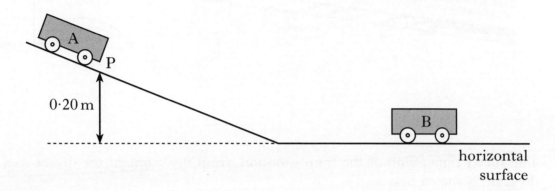

Cart A is released, runs down the slope and collides with cart B. The carts stick together and move off along the horizontal surface.

(a) Calculate the change in gravitational potential energy of cart A from point P to the bottom of the slope. **2**

(b) Assuming no energy losses, show that the speed of cart A at the bottom of the slope is 2·0 m/s. **2**

(c) Calculate the speed of the carts just after the collision. **2**

(d) Describe how the instantaneous speed of the carts immediately after the collision can be measured.

List any apparatus required and state all the measurements that should be taken. **3**

(9)

[Turn over

22. The driver of a train travelling at 45 m/s sees a sign indicating that there is a speed limit of 10 m/s on a bridge on the track ahead. At this point the distance from the train to the bridge is 500 m.

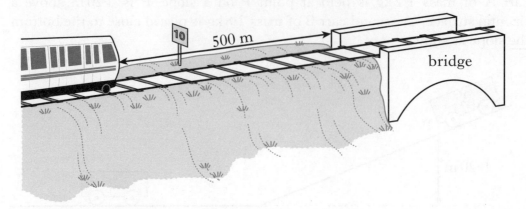

The speed-time graph of the train's motion, from the moment the driver sees the sign, is shown below.

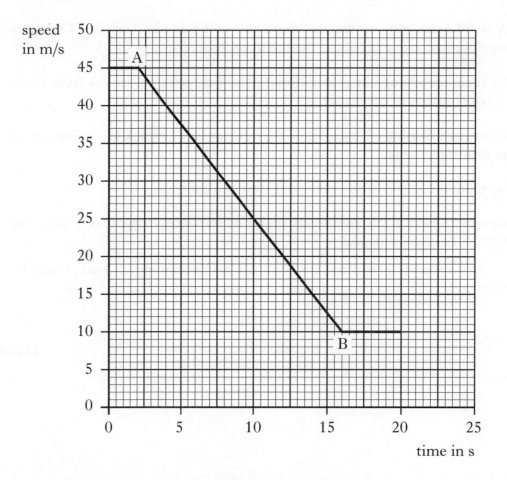

(a) (i) State the time at which the driver starts to apply the brakes.

(ii) Explain your answer.

(b) Calculate the acceleration of the train between A and B.

(c) Is the train travelling at 10 m/s when it reaches the bridge?
You **must** justify your answer by calculation.

23. A boat lift is used to move boats between two canals at different heights as shown.

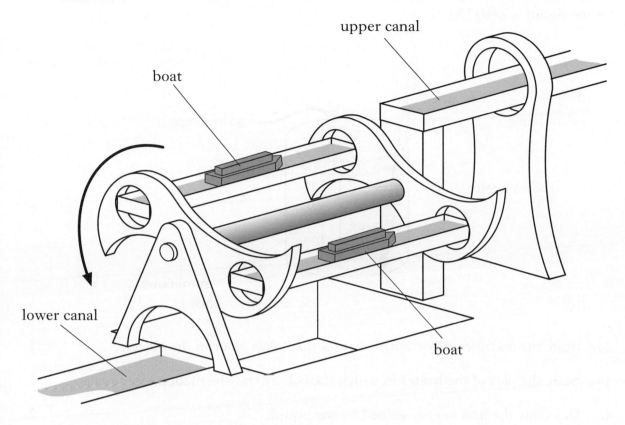

The lift rotates, lowering a boat on one side while raising a boat on the other side.

The system is balanced. The motor rotating the lift has to overcome only frictional forces. Information about the lifting process is shown below.

Frictional force = 84 000 N
Distance through which force acts = 12 m
Time for lifting process to be completed = 4 minutes

(a) Show that the power required to operate the lift is 4·2 kW. 3

(b) The motor operates at 400 V and draws a current of 16 A.

Calculate the input electrical power. 2

(c) Calculate the efficiency of the motor. 2

(d) State whether the power required to start the lift moving is greater than, less than or equal to 4·2 kW. You must explain your answer. 2

(9)

[Turn over

24. A heater immersed in 0·40 kg of a liquid is switched on for 4 minutes. The temperature of the liquid rises by 5 °C in this time. The specific heat capacity of the liquid is 2400 J/kg °C.

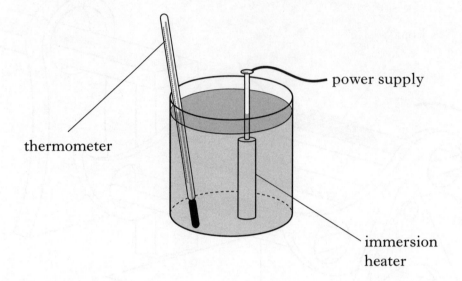

(a) State the useful energy transformation that takes place in the heater. **1**

(b) State the part of the heater in which the energy transformation takes place. **1**

(c) Calculate the heat energy gained by the liquid. **2**

(d) Calculate the power rating of the heater.
State **one** assumption you have made. **3**

(7)

25. A mobile phone charging unit contains a transformer.

The transformer circuit is shown below.

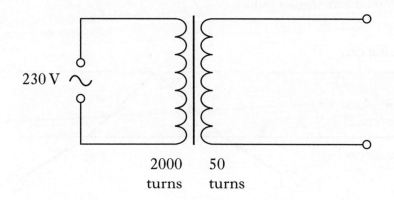

(a) State the purpose of the transformer. **1**

(b) Calculate the secondary voltage of the transformer. **2**

(c) When the phone is being charged the current in the primary coil of the transformer is 24 mA.

Calculate the current in the secondary coil of the transformer. **2**

(d) The mobile phone transmits microwaves at a frequency of 1800 MHz.

Calculate the wavelength of the microwaves. **3**

(8)

[Turn over

26. A car has a system that switches on the windscreen wipers when rain is detected on the windscreen.

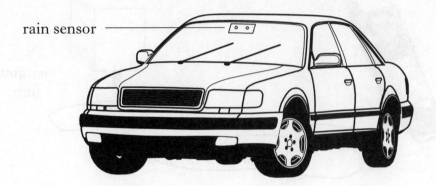

The rain sensor contains an LED which emits a beam of infrared radiation inside the car. In dry conditions this beam travels through the glass and is picked up by a detector as shown below.

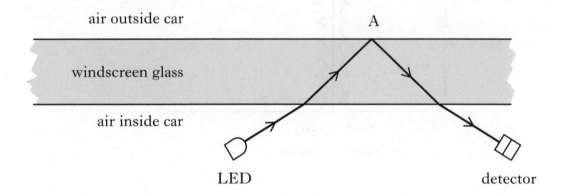

(a) (i) Name the effect on the beam at A. **1**

(ii) Draw the symbol for an LED. **1**

(iii) The LED circuit is shown below.

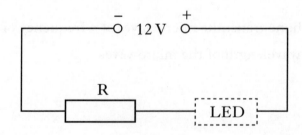

The voltage across the LED is 1·8 V and the current in the LED is 100 mA.

Calculate the resistance of R. **3**

26. (continued)

(b) When rain falls on the windscreen the detector picks up less infrared radiation and the windscreen wipers are switched on.

A student builds a model of the system. The model uses an LDR to represent the infrared detector and visible light to represent infrared radiation. The circuit is shown below.

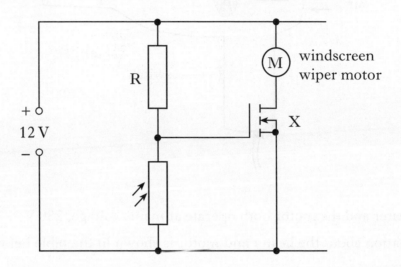

(i) Name component X. **1**

(ii) Describe how this electronic system operates when less light falls on the LDR. **3**

(9)

[Turn over

27. A paint stripper contains a heater and a motor which drives a fan.

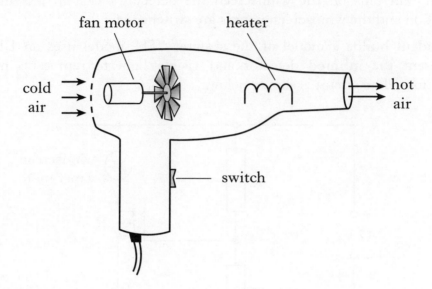

The heater and the motor both operate at mains voltage, 230 V.

Information about the heater and motor is shown in the table below.

	Heater	Motor
Symbol	—⎕⎕⎕—	—(M)—
Operating voltage	230 V	230 V
Power	1425 W	575 W

(a) Calculate the resistance of the motor.   2

(b) Draw the circuit diagram for the paint stripper.   2

(c) The heater burns out. What effect, if any, does this have on the speed of the fan motor?

You **must** explain your answer.   2

(6)

28. Radio waves are transmitted between New York and Edinburgh.

The ionosphere is a layer of charged particles above the Earth.

Radio waves with frequencies below 40 MHz are reflected by the ionosphere.

Radio waves with frequencies above 40 MHz pass through the ionosphere.

diagram not to scale

(a) What is transferred by a radio wave? **1**

(b) An aerial in New York transmits and receives signals of the following frequencies.

    300 kHz   3 MHz   30 MHz   300 MHz

Which of these frequencies could be used for communication with Edinburgh by **satellite**?

You **must** give a reason for your answer. **2**

(c) A satellite is 36 000 km from both transmitting and receiving stations in New York and Edinburgh.

Calculate the minimum time for a signal to pass from New York to Edinburgh using the satellite. **3**

(6)

[Turn over

29. (*a*) An optician wishes to attach labels to spectacles to show the power of their lenses. The following labels are available.

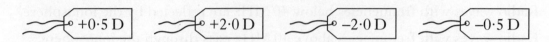

One pair of spectacles has two lenses of the same power. The optician uses one of the lenses to obtain a sharp image of a distant window on a piece of paper as shown below.

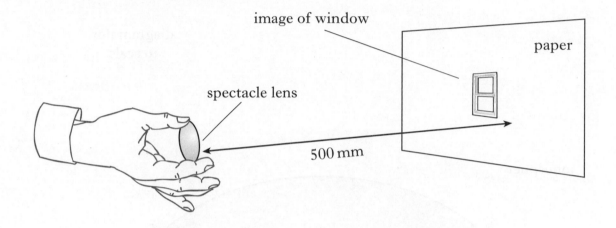

    (i) Name this type of lens. **1**

    (ii) Which label should be attached to these spectacles?

        You **must** show clearly the working which leads to your answer. **3**

(*b*) The diagram below shows a ray of light incident on a glass surface.

Copy the diagram and complete it to show the normal and the refracted ray.

On your diagram, label the angle of incidence *i* and the angle of refraction *r*. **2**

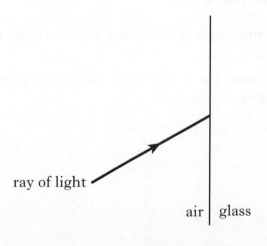

(6)

30. In the reactor of a nuclear power station a uranium nucleus is bombarded by a slow neutron as shown below.

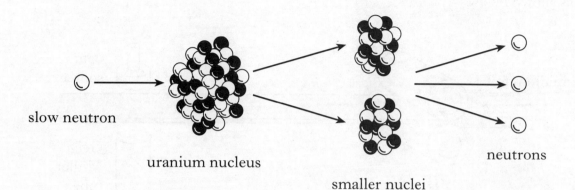

(a) State the name of this type of nuclear reaction. **1**

(b) In this reaction neutrons are released.

Why are these neutrons important to the operation of the reactor? **1**

(c) The reactor also contains boron control rods.

Explain the purpose of these rods. **1**

(d) A worker in the power station is exposed to the following absorbed doses.

2·0 mGy of slow neutrons
5·0 µGy of fast neutrons

The table below gives quality factors of various types of radiations.

Radiation	Quality factor
X-rays	1
gamma rays	1
slow neutrons	3
fast neutrons	10
alpha particles	20

Calculate the total dose equivalent received by the worker. **3**

(e) (i) State **one** advantage of using nuclear power for the generation of electricity. **1**

(ii) State **one** disadvantage of using nuclear power for the generation of electricity. **1**

(8)

[Turn over for Question 31 on *Page twenty*

31. A roller mill produces thin sheets of aluminium foil. The thickness of the foil is checked using a source of beta radiation, a Geiger-Müller tube and a counter as shown below.

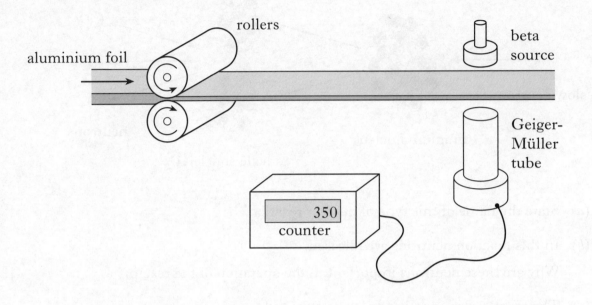

(a) What happens to the count rate when the thickness of the foil increases? **1**

(b) Why is an alpha source **not** suitable for this system? **1**

(c) Radioactive sources give off radiations that cause ionisation.

   (i) What is meant by *ionisation*? **1**

   (ii) Give **two** precautions that should be taken when handling radioactive sources. **2**

(5)

[END OF QUESTION PAPER]

2005 | Intermediate 2

[BLANK PAGE]

X069/201

NATIONAL
QUALIFICATIONS
2005

TUESDAY, 24 MAY
1.00 PM – 3.00 PM

PHYSICS
INTERMEDIATE 2

Read Carefully

1 All questions should be attempted.

Section A (questions 1 to 20)

2 Check that the answer sheet is for Physics Intermediate 2 (Section A).

3 Check that the answer sheet you have been given has **your name**, **date of birth**, **SCN** (Scottish Candidate Number) and **Centre Name** printed on it.

Do not change any of these details.

4 If any of this information is wrong, tell the Invigilator immediately.

5 If this information is correct, **print** your name and seat number in the boxes provided.

6 Use **black** or **blue ink** for your answers. **Do not use red ink**.

7 There is **only one correct** answer to each question.

8 Any rough working should be done on the question paper or the rough working sheet, **not** on your answer sheet.

9 At the end of the exam, put the **answer sheet for Section A inside the front cover of your answer book**.

10 Instructions as to how to record your answers to questions 1–20 are given on page two.

Section B (questions 21 to 31)

11 Answer the questions numbered 21 to 31 in the answer book provided.

12 Fill in the details on the front of the answer book.

13 Enter the question number clearly in the margin of the answer book beside each of your answers to questions 21 to 31.

14 Care should be taken to give an appropriate number of significant figures in the final answers to calculations.

SECTION A

For questions 1 to 20 in this section of the paper the answer to each question is either A, B, C, D or E. Decide what your answer is, then put a horizontal line in the space provided—see the example below.

EXAMPLE

The energy unit measured by the electricity meter in your home is the

    A ampere

    B kilowatt-hour

    C watt

    D coulomb

    E volt.

The correct answer is **B**—kilowatt-hour. The answer **B** has been clearly marked with a horizontal line (see below).

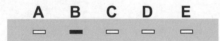

Changing an answer

If you decide to change your answer, cancel your first answer by putting a cross through it (see below) and fill in the answer you want. The answer below has been changed to **B**.

If you then decide to change back to an answer you have already scored out, put a tick (✓) to the **right** of the answer you want, as shown below:

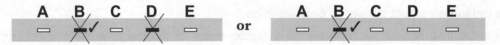

SECTION A

Answer questions 1–20 on the answer sheet.

1. Which of the following is a scalar quantity?

   A Velocity
   B Displacement
   C Acceleration
   D Force
   E Speed

2. At an airport an aircraft moves from the terminal building to the end of the runway.

   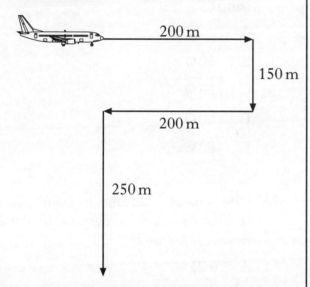

   Which row shows the total distance travelled and the size of the displacement of the aircraft?

	Total distance travelled (m)	Size of displacement (m)
A	400	800
B	450	200
C	450	400
D	800	400
E	800	800

3. A seagull, flying horizontally at 8 m/s, drops a piece of food. What will be the horizontal and vertical speeds of the food when it hits the ground 2·5 s later? Air resistance should be ignored.

	Horizontal speed (m/s)	Vertical speed (m/s)
A	0	8
B	8	20
C	8	25
D	25	25
E	33	50

4. A block of mass 4 kg is pulled along a horizontal bench by a force of 16 N.

   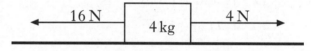

   A constant frictional force of 4 N acts on the block.

   What is the acceleration of the block?

   A $0·3 \text{ m/s}^2$
   B $1·0 \text{ m/s}^2$
   C $3·0 \text{ m/s}^2$
   D $4·0 \text{ m/s}^2$
   E $5·0 \text{ m/s}^2$

   [Turn over

5. Which of the following could be the unit of kinetic energy?

   A N m²
   B N m/s
   C kg m/s
   D N/kg
   E kg m²/s²

6. In observing a collision, a student draws the following conclusions.

   I   Momentum is conserved.
   II Momentum is a vector quantity.
   III Momentum is the product of mass and velocity squared.

   Which of these conclusions is/are correct?

   A I only
   B I and II only
   C I and III only
   D II and III only
   E I, II and III

7. An ampere is one
   A volt per joule
   B joule per second
   C joule per coulomb
   D coulomb per second
   E ohm per volt.

8. The diagram shows a transformer connected to a lamp.

   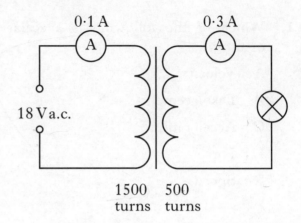

   Assuming the transformer to be 100% efficient, what is the resistance of the lamp?

   A 20 Ω
   B 30 Ω
   C 54 Ω
   D 60 Ω
   E 180 Ω

9. The charge passing a point in a conductor when a current of 4 mA flows for 1000 s is

   A 0·25 C
   B 0·4 C
   C 4 C
   D 250 C
   E 4000 C.

10. Which row shows the frequency and voltage of the mains supply?

	Frequency (Hz)	Quoted voltage (V)	Peak voltage (V)
A	10	110	230
B	50	230	230
C	50	230	325
D	60	230	162
E	230	50	50

11. Which of the following devices converts electrical energy to kinetic energy?

A Motor
B Lamp
C LED
D LDR
E Microphone

12. The diagram shows a transverse wave.

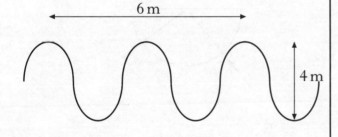

The amplitude of the wave is

A 2 m
B 3 m
C 4 m
D 6 m
E 8 m.

13. Which graph shows how the resistance of most thermistors varies with temperature?

A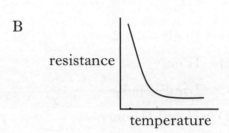

B

C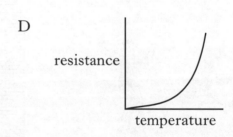

D

E

[Turn over

14. An electric kettle is rated at 2·76 kW for use on a 230 V supply. Which of the following statements is/are correct?

  I The kettle uses energy at the rate of 2·76 joules per second.

  II The current through the element of the kettle is 12 A.

  III 230 coulombs of charge flow through the element every second.

  A I only
  B II only
  C I and II only
  D II and III only
  E I, II and III

15. The diagram shows a girl standing at a fireworks display. There is a tall building nearby.

tall building

When a firework explodes, the girl hears two bangs 0·5 s apart.

The speed of sound is 340 m/s.

How far is the girl from the building?

A 42·5 m
B 85·0 m
C 170 m
D 340 m
E 680 m

16. The diagrams show a light ray passing through a semi-circular glass block.

In each case one angle has been marked.

In which diagram is this angle the critical angle?

A
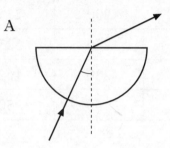

B
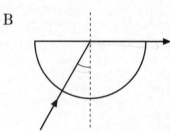

C
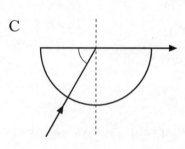

D
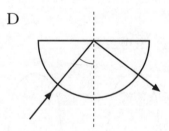

E
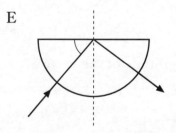

17. Which row correctly describes alpha, beta and gamma radiations?

	α	β	γ
A	electrons from the nucleus	helium nucleus	electro-magnetic radiation
B	electro-magnetic radiation	helium nucleus	electrons from the nucleus
C	helium nucleus	electro-magnetic radiation	electrons from the nucleus
D	helium nucleus	electrons from the nucleus	electro-magnetic radiation
E	electro-magnetic radiation	electrons from the nucleus	helium nucleus

18. Which of the following increases the dose equivalent from a radioactive source?

- A Increasing distance
- B Handling with tongs
- C Standing beside it for a long time
- D Storing in a lead container
- E Storing under water

19. A radioactive source emits α, β and γ radiations in a beam as shown.

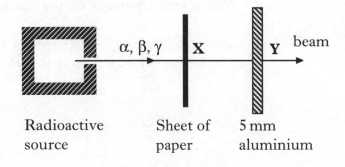

Radioactive source Sheet of paper 5 mm aluminium

The main radiation(s) in the beam at **X** and **Y** are

	Position **X**	Position **Y**
A	α and β	β
B	β and γ	β
C	α and γ	γ
D	α and β	α
E	β and γ	γ

20. The following is an extract from a student's notes on nuclear fission.

  I The nucleus splits into two parts.

  II Neutrons are emitted.

  III Two nuclei join together.

Which of the statements is/are correct?

- A I only
- B II only
- C III only
- D I and II only
- E I, II and III

[Turn over

SECTION B

Marks

Write your answers to questions 21–31 in the answer book.

21. In a game of bowls, a bowler moves a bowl through a horizontal distance of 1·5 m from rest before releasing it with a velocity of 10 m/s.

Mass of bowl = 1·5 kg
Mass of jack = 0·25 kg

(a) Show that the kinetic energy of the bowl when it is released is 75 J. **2**

(b) Calculate the force the bowler applies to the bowl. **2**

(c) The bowl has a speed of 2 m/s when it hits the stationary jack.

After the collision the speed of the bowl is 1·2 m/s.

Calculate the speed of the jack after the collision. **2**

(d) Describe a method to find the average speed of the bowl from the moment it is released until it hits the jack.

Your answer should include:

- the apparatus required
- the measurements taken
- how the average speed is calculated. **3**

(9)

22. A sky-diver of mass 90 kg drops from a stationary balloon. The speed-time graph shows how the vertical speed of the sky-diver varies until she reaches the ground. She falls 3000 m before opening her parachute.

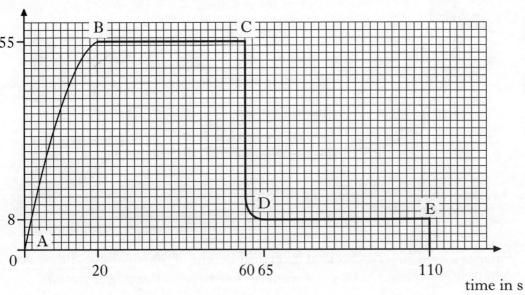

(a) At what point does the sky-diver:

(i) open her parachute; **1**

(ii) reach the ground? **1**

(b) Sketch a diagram showing the forces acting on the sky-diver between B and C.

You must name these forces and show their directions. **2**

(c) Calculate the force of friction acting on the sky-diver between B and C. **3**

(7)

[Turn over

23. A new hydro-electric power station is being planned for the Highlands.

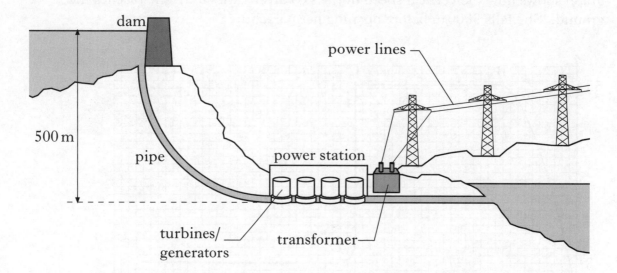

(a) Water is stored in a reservoir at a vertical height of 500 m above the power station. Each second 8000 kg of water flows through the power station.

 (i) Show that the water loses 40 MJ of gravitational potential energy each second. **2**

 (ii) Assuming no energy losses in the pipe, state the input power to the station. **1**

 (iii) Calculate the electrical output power of the station if it is 80% efficient. **2**

(b) The output current from the power station is 1280 A at a voltage of 25 kV.

The voltage is stepped-up to 400 kV by a transformer.

Assuming no energy losses in the transformer, calculate the current in the power lines. **2**

(7)

24. An ice cream maker has a refrigeration unit which can remove heat at 120 J/s. Liquid ice cream, of mass 0·6 kg at a temperature of 20 °C, is added to the container.

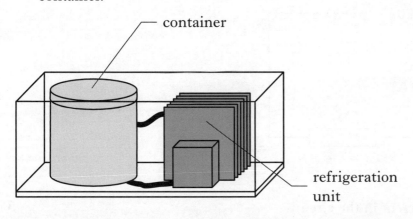

(a) Calculate how much energy must be removed from the mixture to cool it to its freezing point of −16 °C.
(Specific heat capacity of ice cream = 2100 J/kg °C) **2**

(b) Calculate how much heat energy must be removed to freeze the ice cream at this temperature.
(Specific latent heat of fusion of ice cream = $2·34 \times 10^5$ J/kg) **2**

(c) (i) Calculate the time taken to cool and freeze the ice cream. **3**

(ii) What assumption have you made in carrying out this calculation? **1**

(8)

[**Turn over**

25. (*a*) A student connects two resistors in series with a power supply set at 20 V.

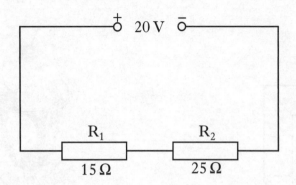

    (i) Calculate the current in the circuit. **3**

    (ii) Calculate the potential difference across resistor R_1. **2**

    (iii) Redraw the above circuit diagram showing meters correctly connected to measure the quantities in (i) and (ii) above. **2**

(*b*) R_1 is now replaced by a 4 V lamp and the supply voltage is reduced to 12 V.

The lamp is operating at its stated voltage.

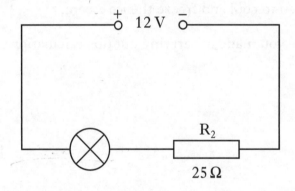

Calculate the rate at which electrical energy is converted to heat energy in resistor R_2. **3**

(10)

26. A student makes a device to measure the speed of moving air.

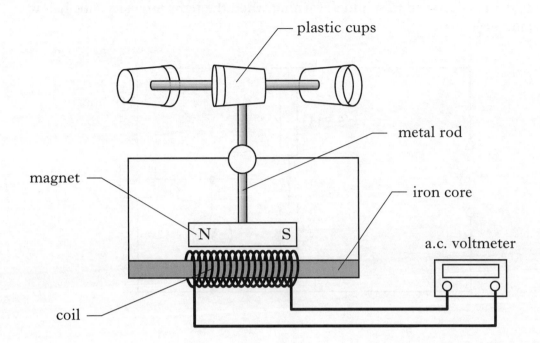

Moving air pushes the plastic cups round and this turns the metal rod.
A bar magnet is attached to the rod as shown in the diagram.

(a) Explain why a reading is shown on the voltmeter when the cups move. 2

(b) What happens to the voltmeter reading when the air speed increases? 1

(c) Suggest **two** changes which could be made **to the apparatus** to give a bigger reading on the voltmeter. 2

(d) Explain, in terms of electron flow, what is meant by a.c. 1

(6)

[Turn over

27. A circuit diagram of an electronic system is shown below.

The system is designed to sound a warning when the light intensity falls below a certain level.

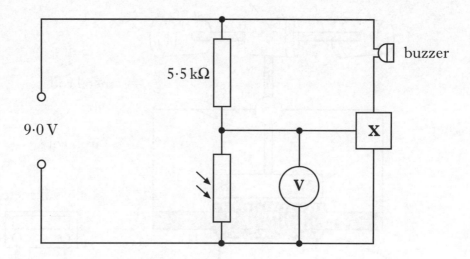

(a) Component X is a transistor.

Two types of transistor are suitable for this system, an NPN transistor and an n-channel enhancement MOSFET.

Draw and name the circuit symbol for each transistor. **2**

(b) What is the purpose of the transistor in this system? **1**

(c) A MOSFET is used at position X.

When the light intensity falls, the voltmeter reading rises to 2·4 V and the buzzer sounds.

Calculate the resistance of the LDR when this happens. **3**

(6)

28. A student is investigating an amplifier system.

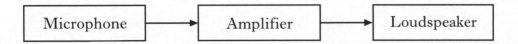

(a) The student whistles into the microphone at a frequency of 400 Hz.

Calculate the wavelength of the sound emitted by the loudspeaker. (The speed of sound in air is 340 m/s.) 2

(b) The student obtains the following information about the system.

Output voltage of the microphone = 2 mV
Resistance of loudspeaker = 16 Ω
Current in loudspeaker = 25 mA

Calculate the voltage gain of the amplifier. 3

(5)

[Turn over

29. The sun is 1.5×10^{11} m from the Earth. The sun emits all radiations in the electromagnetic spectrum. All these radiations travel through space at 3×10^8 m/s.

(a) What do all waves transfer? **1**

(b) Calculate the time taken for sunlight to reach Earth. **2**

(c) The diagram below shows the electromagnetic spectrum in order of increasing frequency.

One part has been missed out.

Radio & TV		Infrared	Visible light	Ultraviolet	X-rays	Gamma rays

(i) Name the missing radiation. **1**

(ii) Name an ionising radiation from the spectrum. **1**

(iii) What is meant by *ionisation*? **1**

(6)

30. (a) An osprey sees a fish in a loch.

The diagram shows the path of a light ray from the fish to the osprey.

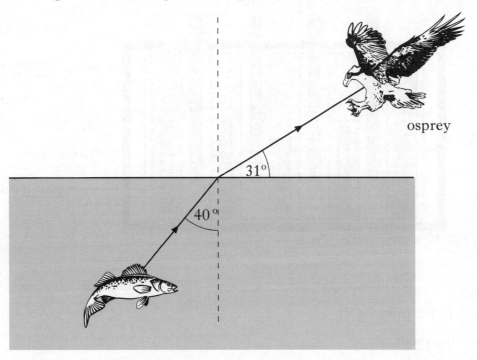

(i) State the size of the angle of incidence. 1

(ii) State the size of the angle of refraction. 1

(b) A bird watcher is using a telescope to watch the osprey.

The eyepiece of the telescope acts as a magnifying lens.

(i) On the square paper provided, copy the diagram below.

Complete your diagram to show the size and position of the image formed by the lens. 2

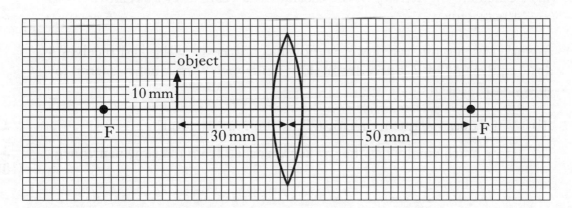

(ii) The focal length of this lens is 50 mm. Calculate the power of this lens. 2

(6)

[Turn over for Question 31 on *Page eighteen*

31. A diagram of the core of a gas cooled nuclear reactor is shown below.

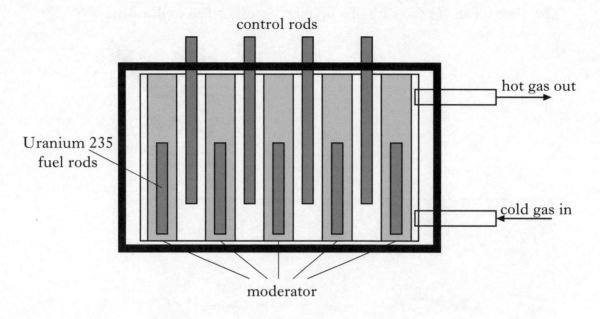

(a) Explain the purpose of

    (i) the moderator 1

    (ii) the control rods. 1

(b) One of the waste products produced in the nuclear reactor is caesium.

The caesium in the waste products removed from the reactor has an activity of 16×10^{12} Bq.

Caesium has a half-life of 30 years.

    (i) State what is meant by the activity of a radioactive source. 1

    (ii) State what is meant by the half-life of a radioactive source. 1

    (iii) Calculate the activity of the caesium 150 years after its removal from the reactor. 2

(c) A worker at the nuclear power station has a mass of 90 kg and receives a dose equivalent of 276 µSv from slow neutrons. The quality factor for slow neutrons is 2·3.

    (i) What does the quality factor tell us about a radiation? 1

    (ii) How much energy has the worker absorbed from the slow neutrons? 3

(10)

[END OF QUESTION PAPER]

[BLANK PAGE]